FUEL CELL ELECTRONICS PACKAGING

FUEL CELL ELECTRONICS PACKAGING

Edited by

Ken Kuang
Torrey Hills Technology, LLC
San Diego, CA 92121

Keith Easler
Kyocera America, Inc.
San Diego, CA 92123

Springer

Editors:
Ken Kuang
Torrey Hills Technologies, LLC
10401 Roselle St., Suite 205
San Diego, CA 92121

Keith Easler
Kyocera America, Inc.
8611 Balboa Avenue
San Diego, CA 92123

Fuel Cell Electronics Packaging

ISBN-13: 978-1-4419-4289-0 e-ISBN-13: 978-0-387-47324-6

Printed on acid-free paper.

9 8 7 6 5 4 3 2 1

springer.com

Foreword

A Look at Fuel Cells from Inside the Beltway

It is an interesting time to be involved with fuel cells.

After President Bush's 2003 State of the Union address, companies involved in fuel cells and hydrogen were riding a wave of national attention. The president's 10-year program was an exciting, long-term effort designed to dramatically change the way Americans use and harness energy. With funding proposed at $1.2 billion for the first five years, it was clear that the federal government was not only making a handsome investment in fuel cells and hydrogen, but also serious about becoming more energy independent.

Despite the fact that the primary focus of the new initiative revolved around automotive technologies, the President's Hydrogen Fuel Initiative was crafted into a balanced program that benefited a wide range of technologies and applications, including micro, portable, stationary fuel cells.

This massive effort was given an additional nod after Congress passed, and the president signed into law, the Energy Policy Act of 2005. The bill called for new levels of research and development, demonstrations, early market incentives, as well as tax credits.

While the president praised the new package, and kept his commitment to the original $1.2 billion, neither he nor the Congress has yet proposed new appropriations for these programs.

Clearly, the industry must make its case anew, to counter critics who question the administration's motives and the potential of the technology, but also to justify program increases in a financially constrained environment.

There is plenty to be optimistic about. First, leading industry officials, as well as the National Academy of Sciences, remain steadfast in their support for fuel cells as the only long-term solution to our energy problems. Second, micro and portable fuel cell products are beginning to emerge. In fact, the US Fuel Cell Council is leading an effort to allow portable fuel cell devices aboard aircraft, in the process removing a significant barrier to consumer adoption.

Also, stationary and larger portable fuel cells have progressed to the point where they are being adopted on their own merits in some high-value markets. Early adopters are also taking advantage of the investment tax credit for fuel cells which were included in the Energy Policy Act. These early deployments have the added benefit of helping to educate the public and state and local code enforcement officials about fuel cells, their fuels and their capabilities.

The US Fuel Cell Council recently released a list of available fuel cell products from our membership. There were more than 40 products on the list – a powerful indication that more and more fuel cells will be making their way into the hands of mainline consumers. Finally, fuel cell producers have also begun making significant investments in manufacturing capabilities, another powerful leading indicator.

That said, fuel cell technologies and the fuel cell industry have a great deal still to do before fuel cells enter the commercial mainstream. Customers need more and better information and regulatory systems need to become more fuel cell friendly. Fuel cell companies must also do a better job of explaining to policy makers and the public what a fuel cell future can bring.

Fuel cells have a lot to offer, and it is about time we let everyone know.

Bud Deflaviis, United States Fuel Council

Acknowledgements

The editors would like to thank the participants in the 2006 IMAPS Fuel Cell Advanced Technical Workshop for their support and dedication to the education of engineers and professionals working in the electronic areas of fuel cell research and development. Without their efforts this book would not have been possible.

We would also like to give a special thanks to Dr Robert Hockaday whose knowledge, dedication, and patience to those willing to learn was paramount for many young engineers starting their paths in fuel cell science and technology.

Also, we would like to thank Dr Jerry Hallmark whose encouragement and dedication to the proliferation of fuel cells establishes their wide spread introduction.

Finally, we would like to thank Mike O'Neil, Brian Schieman, Michael O'Donoghue, and many others from the IMAPS organization for believing in us and our drive to introduce fuel cell technologies to our colleagues for the enrichment of all who use them.

Contents

Chapter 1

INTRODUCTION TO FUEL CELL TECHNOLOGY

Áron Varga[1]
Department of Materials Science and Engineering, MIT, Cambridge, MA 02139, USA

INTRODUCTION

Fuel cells are electrochemical devices that convert chemical energy directly into electrical energy. All types are based on the same principles and have four basic components: electrolyte, anode, cathode, and interconnect. The different fuel cell types are categorized according to their electrolyte since it determines key features such as operating temperature, power-up time, type of fuel, migrating ions, and shock resistance. Fuel cells are generally utilized for secondary power generation, since in cases where they are not using fossil fuels – a possibility only for high temperature fuel cells – pure hydrogen has to be generated by using primary energy sources. Working fuel cell systems have already been developed by many companies in the automobile, electronics, and power generation industries. These systems have to be improved before they can compete on the market with existing technologies, by extending their lifetime and significantly reducing their cost. Two high temperature (solid oxide and molten carbonate) and two low temperature (polymer electrolyte and direct methanol) fuel cells are discussed in more detail, with a focus on the materials and the electrochemical reaction.

At this point, the quest for alternative energy sources has to proceed without focusing on a single energy technology, but rather following a parallel strategy and letting the new technologies compete before widespread replacement of fossil fuels.

FUEL CELLS IN CONTEXT

The world hunger for energy is increasing more or less in correlation to economic growth. In the past decades, the global community witnessed unprecedented economic development in many regions, such as Southeast

---

[1] Permanent Address: Department of Material Science and Metallurgy, University of Cambridge, Pembroke Street, Cambridge CB2 3QZ, UK.

Asia, Eastern Europe, and the Middle East. Rapid economic growth brought a broad rise in living standards of large populations and with that a steep increase in world energy consumption.

Today's energy demand (Figure 1-1) is met largely by fossil fuels, such as oil, gas, and coal. Only 18% of primary energy sources were renewable in 2003, of which 16% was hydropower. The remaining portion comes from combustible and renewable waste with only a minor contribution from solar, wind, and geothermal energy sources.

Because of their wide availability, their easy accessibility, and their abundance, carbon-based fuels tend to be the energy source of choice. Renewables will be competitive in the medium term only if significant scientific and technological breakthroughs are achieved, unpriced external consequences of burning carbon based fuels are taken into account, and a carbon tax is introduced. Such external consequences include global warming, leading to coral bleaching, melting of glaciers and ice caps, rise in sea level, storms, and flooding, etc. The severity of the consequences do not allow for a "wait and see" approach, but without policy incentives the current socioeconomic inertia will not be overcome and the development of new technologies will not come soon enough. It is of utmost importance to support all currently available alternative energy technologies with a parallel policy approach, since their suitability depends heavily on the application and the region. In addition to the primary alternative energy sources of hydroelectric, geothermal, ocean/tides, wind, biomass, and solar – for which new and better energy storage systems have to be developed – secondary power generation sources must also be pursued.

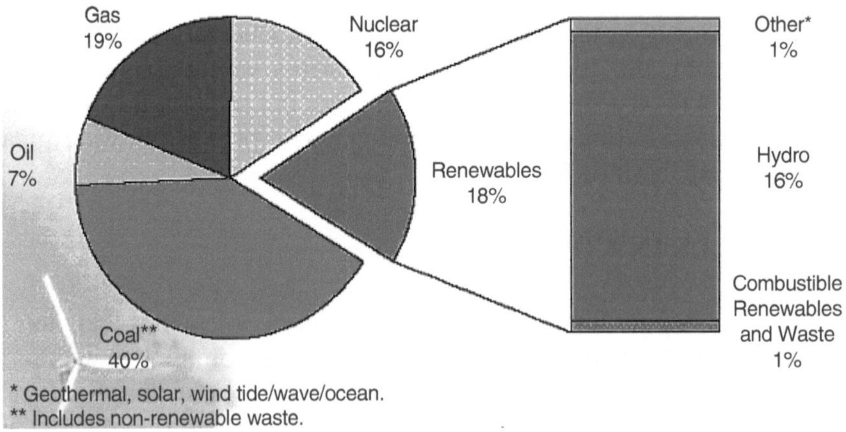

Figure 1-1. Renewables in electricity production [1]. Copyright © OECD/IEA

Fuel cell technology is a very promising potential candidate to help replace the hydrocarbon economy with an alternative economy, perhaps based on hydrogen. A major advantage of the hydrogen economy is its similar structure to the hydrocarbon economy with its possible centralized layout.

As the field of fuel cell research and business expands, there is an increasing necessity to provide educational and background resources. Fuel cells are secondary energy sources since their principal fuel, hydrogen (ignoring for the moment natural gas and other hydrocarbons that are used exclusively for high temperature fuel cells), has to be generated using primary energy sources, such as nuclear, solar, or wind energy. Fuel cells could provide a means to render power generation more efficient, and a relocation of poisonous exhaust gas emissions could be achieved from urban to rural areas, where they would affect fewer people. Fuel cells are suitable for many different applications. Because of their high and constant efficiency even at varying or low load, they cause negligible air pollution (if fossil fuels are used; otherwise none). They are quiet or completely silent and minimize maintenance costs, since no or very few moving parts are used. These advantages are summarized in Table 1-1.

This chapter was intended to provide a summary of four of the most popular fuel cell types: solid oxide, molten carbonate, proton-exchange membrane, and direct methanol fuel cells (DMFCs), with a focus on materials, mechanism of operation, and an evaluation of advantages and disadvantages. William R. Grove, the fuel cell pioneer could only harvest a very small current from his apparatus that was not enough to provide power for everyday use. Only in the last few decades there have been major breakthrough improvements in the design and materials of fuel cells that the technology is now approaching commercialization, attracting increasing attention from the world of business rather than just academia. These improvements are reviewed here before the succeeding chapters take the reader to the cutting edge of today's research.

Table 1-1. The advantages of fuel cells

- Negligible air pollution (if fossil fuels are used; otherwise none)
- Reduced weight, especially in mobile applications
- 100% theoretical efficiency, 80% efficiency in high temperature turbine hybrid systems that can utilize the generated heat
- High efficiency in low power systems
- Constant efficiency even at low load
- Flexible output with rapid adjustment
- Low maintenance cost and very few moving parts (or none)
- Quiet or completely silent

HISTORY AND PRINCIPLES

A fuel cell is a device that converts chemical energy directly into electrical energy, and its operation can be explained using electrochemical concepts that are well understood. The simplest form was developed in 1839 by the talented British chemist William R. Grove, using a beaker, two test tubes, platinum coated wires, and a dilute acid.

By connecting the wires in Figure 1-2 to a power source, water is electrolyzed. The oxygen and hydrogen gases generated are captured in the test tubes that are physically separated. After the power source is disconnected, the cell (or gas battery), as it was first known, runs in reverse: In the presence of platinum electrodes, which are necessary as catalysts, the electrolysis will essentially run backward, and current can be made to flow through a circuit between the two electrodes since hydrogen and oxygen have a natural tendency to react and form water. Their combined free energy in the form of a water molecule is lower than the sum of the pure components, hence in the presence of enough activation energy the direct, exothermic, and explosive reaction of hydrogen and oxygen – the combustion reaction (1) – occurs and must be prevented by the gas impermeable electrolyte in order to harness the energy for useful purposes.

$$H_2 + \frac{1}{2} O_2 \longleftrightarrow H_2O \qquad (1)$$

In the combustion reaction, oxygen is reduced and fuel is oxidized, with the release of heat. The fuel cell separates hydrogen and oxygen with

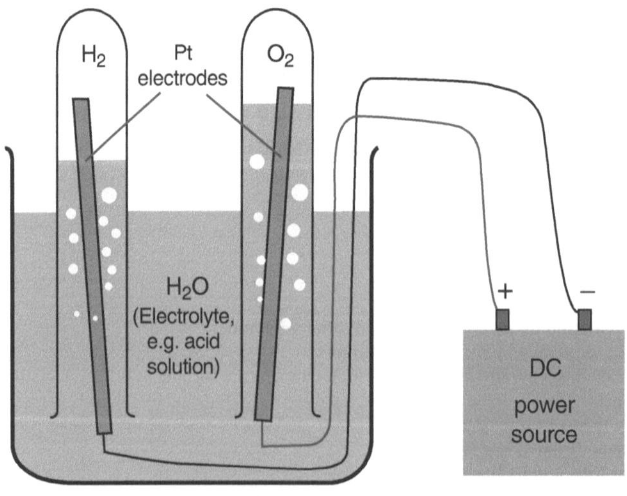

Figure 1-2. Electrolysis setup

an electrolyte through which only ions (e.g., H^+, O^{2-}, CO_3^{2-}) can migrate. Two half reactions, whose character and stoichiometry are determined by the electrolyte, occur at the electrodes. Since ions are readily transported through the electrolyte, oxygen can react only in its reduced form with hydrogen in its oxidized form. The excess electrons from the cathode (reduction side) are transported to the anode (oxidation side). No one tried to make use of the concept demonstrated by Grove until 1889 when Langer and Mond [2] attempted to engineer a practical cell fueled by coal gas. Further early efforts carried on into the early 1900s, but the development of the internal combustion engine made more research into the technology sadly unnecessary.

Francis Bacon developed the first successful fuel cell in 1932, running on pure oxygen and hydrogen and using an alkaline electrolyte with nickel electrodes. It was not until 1959 that Bacon and his colleagues first demonstrated a 5 kW device; the 27 year delay is perhaps an indication of just how difficult it is to make progress in this field of development. Harry Karl Ihrig demonstrated a 20 bhp fuel cell tractor in the same year. Around about this time, NASA started researching the technology, with a view to produce a compact electricity generator for use on spacecraft. Due to their (appropriately!) astronomical budget, it was not long before they succeeded. The Gemini program used early proton exchange membrane fuel cells (PEMFCs) in its later missions, and the Apollo program employed alkaline fuel cells (AFCs). On the spacecraft, the water produced by the reaction was available for the astronauts to drink. Recent developments have come thick and fast as the technology begins to come to fruition. Automotive applications are high on the agenda due to the huge consumer market and the need for an environmentally friendly, renewable alternative to the internal combustion engine and fossil fuels.

Today's fuel cells are very similar to Grove's original design in having only three components serving the same functions: a cathode, an anode, and an electrolyte between the two electrodes. To reach a high enough reaction rate and hence larger current and power output, the electrodes are flat, in order to maximize the contact area with the reactants, and the electrolyte is thin, to minimize the resistance to the flow of ions. The electrodes are designed to be porous to further increase the reaction surface, and catalysts are sometimes added. The most common design is the flat plate, sandwich-like design shown in Figure 1-3.

The applications of fuel cells cover a wide range of possibilities, from mobile systems such as cars and electronic devices to stationary power generation for households and public buildings. The suitability of a fuel cell for a specific application is determined by the electrolyte, since electrolyte properties also specify most of the crucial system characteristics. Therefore

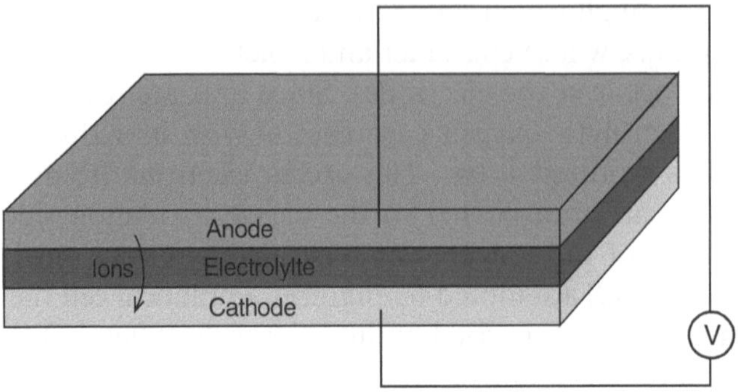

Figure 1-3. Flat plate design for a fuel cell

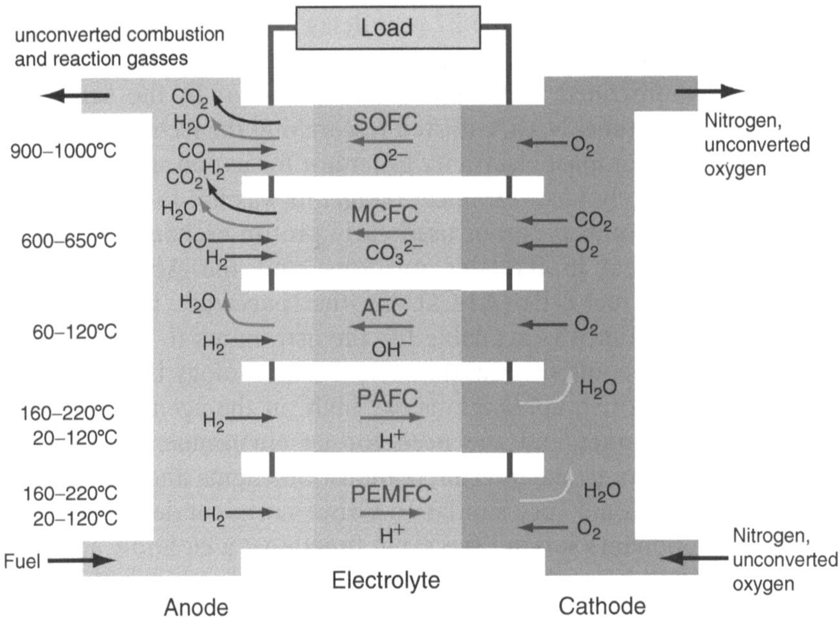

Figure 1-4. Overview of the fuel cell types (after Bewag Vattenfall open source literature)

fuel cells are categorized according to their type of electrolyte; see Figure 1-4 for an outline of the six main types. Polymer electrolyte fuel cells are widely used for mobile applications mainly because of their low operating temperatures. Ceramic electrolyte and molten salt electrolyte fuel cells are restricted to stationary applications, largely because of their high operating temperatures and other material characteristic properties. The reasons for the different application types are discussed in more detail later.

The following sections consider the four most popular types of fuel cells: solid oxide, molten carbonate, proton exchange membrane, and direct methanol. The net reactions are summarized in Tables 2–6 to give the reader an overview.

Grove's fuel cell is one of the simplest examples, the reactions for which are shown in Table 1-2. They are the same for polymer electrolyte fuel cells.

DMFC operates with a proton conducting polymeric electrolyte too, giving the reactions as shown in Table 1-3.

The second most common electrolytes, found mostly in solid oxide fuel cells (SOFCs), are permeable to oxide ions and lead to the half-reactions in Table 1-4.

A third type of electrolyte, used for molten carbonate fuel cells at high temperatures, conducts carbide ions (CO_3^{2-}) and is associated with the electrode reactions in Table 1-5.

One also often encounters alkaline electrolytes, across which OH^- is the transported species. In this case the half-reactions are those listed in Table 1-6.

Table 1-2. Electrode reactions for Grove's fuel cell

Half-reaction at the anode:

$$H_2 \longleftrightarrow 2H^+ + 2e^- \quad (2)$$

Half-reaction at the cathode:

$$O_2 + 4e^- + 4H^+ \longleftrightarrow 2H_2O \quad (3)$$

The net reaction is the combustion reaction:

$$H_2 + \frac{1}{2}O_2 \longleftrightarrow H_2O \quad (1)$$

Table 1-3. Electrode reactions for direct methanol fuel cell

Half-reaction at the anode:

$$CH_3OH + H_2O \longleftrightarrow CO_2 + 6H^+ + 6e^- \quad (4)$$

Half-reaction at the cathode:

$$O_2 + 4H^+ + 4e^- \longleftrightarrow 2H_2O \quad (3)$$

The net reaction is:

$$2CH_3OH + 3O_2 \longleftrightarrow 2CO_2 + 4H_2O \quad (5)$$

Table 1-4. Electrode reactions for SOFCs

Half-reaction at the anode:

$$H_2 + O^{2-} \longleftrightarrow H_2O + 2e^- \quad (6)$$

Half-reaction at the cathode:

$$O_2 + 4e^- \longleftrightarrow 2O^{2-} \quad (7)$$

The net reaction is the same as before:

$$H_2 + \frac{1}{2}O_2 \longleftrightarrow H_2O \quad (1)$$

Table 1-5. Electrode reactions for molten carbonate fuel cells

Half-reaction at the anode:	
$H_2 + CO_3^{2-} \longleftrightarrow H_2O + CO_2 + 2e^-$	(8)
Half-reaction at the cathode:	
$\frac{1}{2}O_2 + CO_2 + 2e^- \longleftrightarrow CO_3^{2-}$	(9)
The net reaction is the combustion reaction:	
$H_2 + \frac{1}{2}O_2 \longleftrightarrow H_2O$	(1)

Table 1-6. Electrode reactions for AFCs

Half-reaction at the anode:	
$H_2 + 2OH^- \longleftrightarrow 2H_2O + 2e^-$	(10)
Half-reaction at the cathode:	
$O_2 + 4e^- + 2H_2O \longleftrightarrow 4OH^-$	(11)
The net reaction is the combustion reaction:	
$H_2 + \frac{1}{2}O_2 \longleftrightarrow H_2O$	(1)

SOLID OXIDE FUEL CELLS

General

In the late nineteenth century, conduction was not yet understood. After major and rapid advances in the field of materials science, Nernst [3] observed at the University of Göttingen that stabilized zirconia (ZrO_2 doped with Ca, Mg, or Y) was an insulator at room temperature, an ionic conductor from 600–1,000°C and a mixed conductor (both electronic and ionic) at around 1,500°C. The main part of the SOFC was therefore discovered. The SOFC concept was demonstrated by Baur and Preis [4] in the 1930s using zirconium oxide, but many improvements were necessary to make a competitive and reliable device. In the 1950s, a simple and straightforward design made cheaper manufacturing processes, such as tape casting, possible: the flat plate fuel cell. SOFCs are the most efficient devices yet invented that can convert chemical energy into electrical energy. Both electrodes (cathode and anode) are made of sophisticated ceramic materials, since the high operating temperature determined by the ceramic electrolyte prevents the use of cheaper, less refractory metals. The principle of operation of the SOFC is straightforward: oxygen atoms are reduced on the porous cathode surface by electrons. The resulting oxide ions diffuse through the electrolyte to the fuel-rich and porous anode, where they react with the fuel (hydrogen and/or carbon monoxide) and give off electrons to an external circuit, as illustrated in Figure 1-5. The exact reaction mechanism is not yet fully understood and requires further investigation.

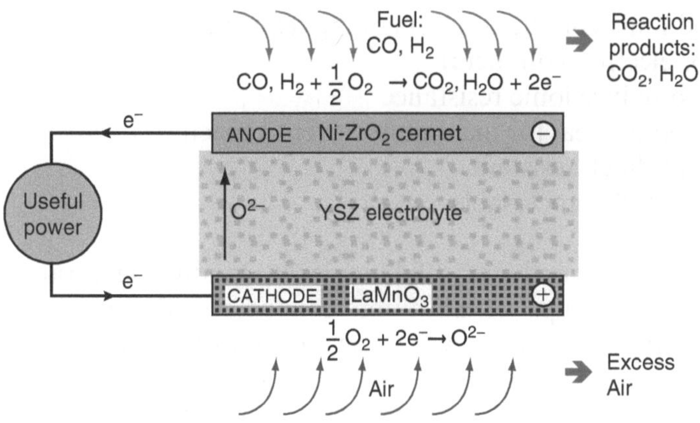

Figure 1-5. Operation of an SOFC

During the operation of the cell, a large amount of heat is produced by the electrochemical reaction, which does not have to be wasted: it can be utilized by an integrated heat management system. Since it takes a long time to reach its operating temperature, the best applications for SOFCs are ones that use both the heat and electricity generated, such as stationary power plants and auxiliary power supplies. Problems with start-up time could be solved by using supercapacitor batteries for the first few minutes of operation in mobile applications. There are a few challenges with the flat plate design when used for larger devices: sealing around the edges, thermal expansion mismatch, and cracking upon thermal cycling, because intrinsically brittle ceramic components are used. Tubular designs have been developed to solve these problems. The big advantage of the SOFC over the molten carbonate fuel cell (MCFC) is the solid nature of the electrolyte, so that there are no pumps required to circulate a hot and corrosive melt. The operating temperature is between 600 and 1,000°C, depending on the generation of the fuel cell (first, second, and third, with, respectively, decreasing operating temperatures). Both hydrogen and carbon monoxide serve as direct fuels, but common hydrocarbon fuels (diesel, natural gas, gasoline, alcohol, etc.) can be supplied to an SOFC if they are reformed internally or externally.

Electrolyte

There are several criteria that the electrolyte has to meet, not only in solid oxide but also in *all* fuel cells. It must be:

- Dense and leak tight
- Stable in reducing and oxidizing environments

- A good ionic conductor at operating temperatures
- A poor electron conductor
- Thin, to reduce ionic resistance
- Extended in area, for maximum current capacity
- Thermal shock resistant
- Economically processable

The materials used are solid, ionically conducting ceramics. There are two main groups of such ion conductors: fluorite-structure and perovskite-structure oxides. A few new materials have emerged, which are hexagonal-structured oxides. The most widely used material is yttria-stabilized zirconia (YSZ); next popular electrolyte materials are doped ceria, CeO_2, and doped lanthanum gallate, $LaGaO_3$ (both are oxygen ion conductors); and doped barium zirconate, $BaZrO_3$ (a proton conductor). The concentration and type of dopants influence the material properties strongly; ionic radius and ion charge are the principle characteristics. Dopants that cause the least strain and hence have the least influence on the potential energy landscape of the parent lattice have the biggest effect on the conductivity. In an oxide ion conductor, current flows by the movement of oxide ions through the crystal lattice. This is a thermally activated process, where the ions hop from one site to another (from one potential valley to another) in a random way. When an electric field is applied, there is a drift in one direction superimposed on the random thermal motion. Ionic conduction depends on the mobility of the ions and therefore on temperature. At high temperatures (600–1,000°C), the conductivity can reach 100 S m^{-1}, which is of the same order of magnitude as for liquid electrolytes. For transport by a vacancy diffusion mechanism, the crystal has to contain unoccupied oxygen ion sites. The energy barrier for ion migration from an occupied site to an unoccupied site must be small (≤1 eV). Oxygen ion migration might still seem unlikely, since the relative size of the oxygen ions is large and the smaller metal ions are more likely candidates to migrate in an electric field. Indeed, only a limited number of materials, such as fluorite-structure oxides, perovskites, the LaMeOx (LAMOX) and BiMeVOx (BIMEVOXes) families facilitate fast oxygen ion migration.

Fluorite oxides are the most common and classical oxygen ion-conducting materials. The crystal structure consists of a cubic oxygen lattice with alternate body centers occupied by 8-coordinated cations. The cations are arranged into a face-centered cubic structure with the anions occupying the tetrahedral sites. This leaves a rather open architecture, with large octahedral interstitial voids, as shown in Figure 1-6.

The general formula has the form AO_2, where A is usually a large tetravalent cation, e.g., U, Th, Ce. Since Zr^{4+} is too small to sustain the

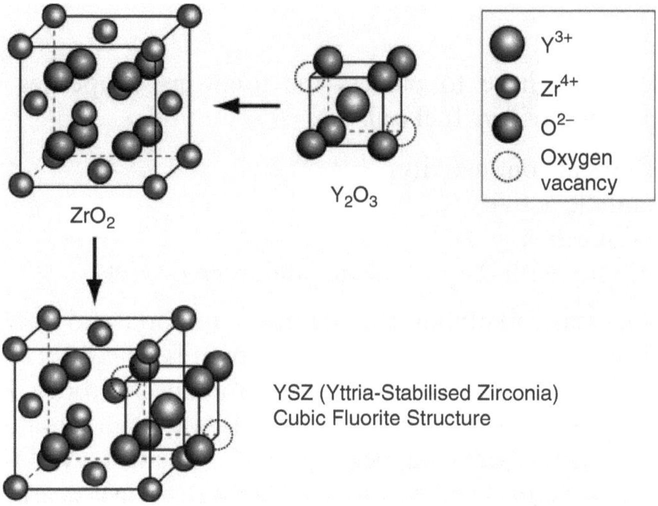

Figure 1-6. Structure of YSZ (created with CrystalMaker™)

fluorite structure at low temperatures, it has to be partly substituted with a larger dopant cation. Doping usually involves substituting lower valence cations into the lattice. In order to maintain charge neutrality, oxygen vacancies have to be introduced, which allow and facilitate oxygen ion migration. An interesting feature of the fluorite lattice is that it can sustain a high degree of substitution. A very disordered and open structured material results from this, promoting ionic conduction.

In present fuel cells, the electrolyte of choice is zirconia, partially stabilized by either 3 mol% Y_2O_3 (3YSZ) or 8 mol% Y_2O_3 (8YSZ). YSZ is not the best ion conductor, but it is the cheapest to process and has low enough electronic conductivity and high enough ionic conductivity. There are many other materials that conduct oxygen ions but the advantages of YSZ (abundance, chemical stability, nontoxicity, and economics) make it the most suitable material at present. One drawback is the high thermal expansion coefficient, which presents problems in sealing the fuel cell. The world demand for YSZ is rising, but fortunately, Zr is one of the most common elements in the Earth's crust usually occurring in the form of the silicate zircon ($ZrSiO_4$). This material has to be converted to zirconia (ZrO_2), since the highly covalent Si ions tend to block ionic and electron paths. Yttria is the main stabilizer used, and about 13–16 wt% have to be added to give fully stabilized cubic material. The supply of more rare dopants, such as scandia, could present a problem in the future. Fabrication of zirconia electrolyte films is usually done by tape casting or vapor deposition.

Electrodes

The electrodes have to possess the following properties to ensure smooth operation of *any* fuel cell:

- High electrical conductivity
- High catalytic activity
- High surface area
- Compatibility with the electrolyte (and interconnect)

For many years, platinum was the most common and only electrode material. The biggest disadvantage of platinum is its cost: it is economically not viable to build fuel cells with platinum electrodes. Hence, the pressure to find cheaper electrode materials was high, resulting in the exploration of perovskite ceramics (e.g., $LaCoO_3$ and $LaMnO_3$). Further development of perovskites was slowed, because of fast degradation due to their reactivity with YSZ. We will further inspect $La_{0.8}Sr_{0.2}MnO_3$ (LSM), which has become a new standard.

The *cathode* material is very important because usually the oxidation reaction determines the efficiency of the fuel cell. Since cathodes operate in a highly oxidizing environment, it is impossible to use cheap base metals. The best compromise is semiconducting oxides, such as doped lanthanum cobaltites and lanthanum manganites. A new material: LSM has suitable properties: good electronic conductivity and matching heat expansion coefficient. For SOFCs whose operating temperature has been reduced to below 1,000°C, it is possible to use LSM mixed with YSZ in equimolar proportions to form the first surface on the electrolyte. Other materials have also been used, such as $La_{0.6}Sr_{0.4}Co_{0.2}Fe_{0.8}O_3$ (LSCF). LSCF has the advantage of lower power losses at lower temperatures and less susceptibility to poisoning by chromium, and it is used with ceria-based electrolytes. The loss of cathode performance is mainly due to changes in the microstructure and the phase composition of the material at load conditions. If stainless steel interconnects are used, degradation caused by Cr poisoning occurs, since it evaporates from the steel and condenses preferably on the cathode. In general, the rate at which Cr poisoning occurs decreases with decreasing operating temperature.

Nickel is used as the *anode* because it is economical and exhibits high performance, although due to reasons of adherence and different expansion coefficients, it flakes off easily from the electrolyte unless it is mixed with zirconia, creating a cermet. Ni–YSZ anodes allow a rapid and clean connection to the fuel and are good electronic conductors although Ni is susceptible to become coated with a carbon layer when it reacts with carbon-based fuels. Coke formation usually impedes further reaction from happening but certain additives to the Ni + YSZ cermet, such as 5% ceria or 1% molybdena inhibit carbon deposition. Besides catalyzing the oxidation of hydrogen, Ni is also active in reforming carbon-containing fuels. Anodes completely made of ceramics show good oxidation–reduction cyclability. The contact sides of the electrolyte with the fuel and the oxidant are coated with the electrode material. To form *porous* contact layers, partially sintered materials are used; and to allow a gradient of properties, such as expansion coefficient, several layers with different compositions are laid down. Since the gas atoms discharge (or absorb) electrons at the anode (or cathode), a three-phase boundary zone is required:

- Gas phase (high porosity electrolyte required for better access)
- Electrolyte phase for ion transport
- Metal phase for electron conduction

Hence a volumetric three-dimensional region has to be provided for the reaction. The most common methods of applying the electrode layer on the electrolyte are plasma spraying, vapor deposition, solution coating and colloidal ink methods.

Interconnection

For the interconnection, an inert and impervious material is needed. It should withstand both oxidizing and reducing environments. Lanthanum chromite ($LaCrO_3$) seems to have the necessary properties in systems operating at up to 1,000°C. Depending on doping, this material matches the thermal expansion coefficient of LSM. For lower temperatures, metallic-based alloys can be used. Again, plasma spraying is the most economic method of applying the interconnect layer onto the electrode. Although lanthanum chromite provides cell life times of up to 70,000 h, it is not perfectly inert: it expands in the presence of hydrogen, causing cracking, especially at large planar stacks.

System and Outlook

A typical SOFC power plant today is fueled with natural gas because of the lack of a hydrogen infrastructure. A plant must have three main components:

1. The *preheater* raises the temperature of the fuel and air to near the operating temperature. At the same time, the preheater reforms the incoming gas mixture, mostly reforming steam to hydrogen. Steam reforming comprises two steps (Table 1-7 and Figure 1-7):

Table 1-7. Preheater reactions

Methane reforming:		
	$CH_4 + H_2O \rightarrow CO + 3H_2$	(12)
Water gas shift:		
	$CO + H_2O \rightarrow CO_2 + H_2$	(13)
Overall reaction:		
	$CH_4 + 2H_2O \rightarrow CO_2 + 4H_2$	(14)

2. The *cell stack* electrochemically oxidizes the hydrogen stream, drawing oxide ions through the electrolyte from the air stream.

$$\text{Electrochemical reaction: } H_2 + \frac{1}{2}O_2 \rightarrow H_2O \quad (1)$$

3. The *lower cycle* utilizes the exhaust heat. The exhaust gases are so hot that gas turbines can be driven to generate additional electrical energy and thus increase the efficiency of the fuel cell system up to 80%. The durability of the SOFC is mainly determined by the processes occurring during thermal cycles, oxidation–reduction cycles and sulfur contamination (even at high temperatures, sulfur is absorbed by the anode).

Due to the high operating temperature, one of the great advantages of the SOFC is that it can use a wide range of fuels, although dependent on the cathode composition. Internal reforming can take place at the anode, when steam is added to the fuel. The reaction of methane is as follows:

$$CH_4 + H_2O \rightarrow CO + 3H_2 \quad (12)$$

Both hydrogen and carbon monoxide can react with the oxide ions. The shift reaction also occurs at the anode, since the reaction of CO is slow, producing more hydrogen.

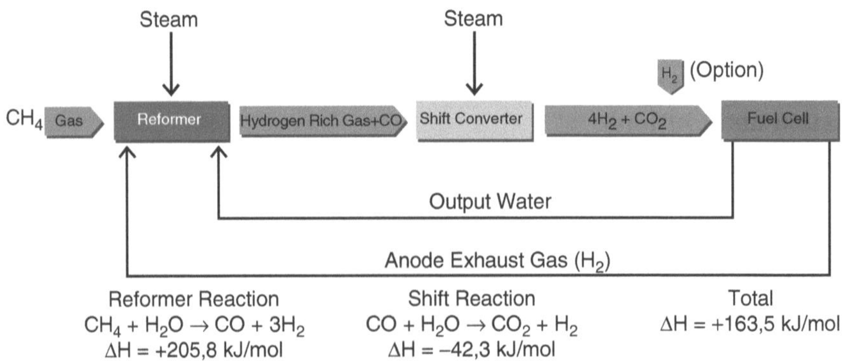

Figure 1-7. Internal reforming in high temperature fuel cells (adapted from Bewag, Vattenfall, open source literature)

$$CO + H_2O \rightarrow CO_2 + H_2 \quad (13)$$

The disadvantage of using hydrocarbon fuels is the possible formation of coke on the anode:

$$2CO \rightarrow CO_2 + C \quad (15)$$

$$CH_4 \rightarrow 2H_2 + C \quad (16)$$

As mentioned above, impurities, such as sulfur are also damaging to the SOFC. Only desulfurized natural gas can be used as fuel. Other additives (more than 100 different molecules are present in commercial gasoline) can have damaging effects on the nickel anode.

The activity of the nickel anode decreases due to sintering and coke formation when carbon-containing fuels are used. The ceramic parts can easily break if vibrational forces are present. This is one reason why SOFCs are best suited for stationary applications rather than mobile applications.

The ultimate goal is to build a decentralized network of medium sized power generating SOFCs that can supply electricity to a small community with much higher reliability and only minor consequences in case of failure compared to the current system of few but very large power plants.

MOLTEN CARBONATE FUEL CELLS

MCFCs are the second type of high temperature fuel cells we are discussing. A molten mixture of salts comprising lithium, sodium, and potassium carbonates is used as the electrolyte. These salts melt and conduct carbonate ions (CO_3^{2-}) from the anode to the cathode when heated to about 600°C, as shown in Figure 1-8. Hydrocarbons have to be used as part of the fuel since the charge carriers in the electrolyte – carbonate ions – have to be supplied. Hydrogen, which can be generated internally by reforming hydrocarbon-based fuels, is also needed at the anode. The electrodes should therefore be resistant to poisoning by carbon. The high exhaust temperature makes cogeneration of electricity with turbines possible; hence the efficiency – 60% without and 80% with hybrid technology – is relatively high compared to other fuel cell systems. MCFCs are mainly utilized for stationary power generation in the 50 kW to 5 MW range. Since it uses a liquid high temperature electrolyte, it is rather unsuitable for mobile applications. The main problem with MCFCs is the slow dissolution of the cathode into the electrolyte. Most of the research is therefore in the area of more durable materials and cathodes.

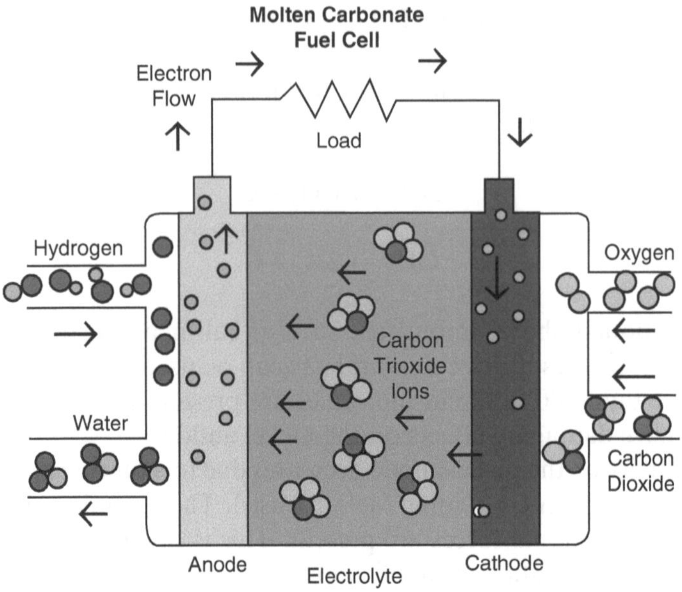

Figure 1-8. Operation of an MCFC

Both SOFC and MCFC are high temperature devices. Their development followed similar lines until the late 1950s. First, Baur and Preis [4] experimented with solid oxide electrolytes in Switzerland. The technical problems they encountered were again tackled by the Russian scientist O.K. Davtyan without success though. In the late 1950s, Dutch scientists G.H.J. Broers and J.A.A. Ketelaar focused on molten carbonate salts as electrolyte and, by 1960, they reported the first MCFC prototype. In the mid-1960s, the US Army's Mobility Equipment Research and Development Center (MERDC) tested several MCFCs made by Texas Instruments, ranging from 100 to 1,000 W. Ishikawajima Heavy Industries showed in Japan in the early 1990s that a 1,000 W MCFC power generator can operate for 10,000 h continuously. Other large power plants with outputs of up to 3 MW are already planned. The MCFC has been under development for 15 years as a stationary electric power plant. But when most problems with SOFC are solved, further work on the MCFC might stop.

Electrolyte

In most cases, the electrolyte of the MCFC is made of a lithium–potassium carbonate salt heated to 600–1,000°C. At this temperature, the salt is in the liquid phase and can conduct ions between the two electrodes. The typical mixture ratio of the electrolyte is 62 mol% Li_2CO_3 and

38 mol% K_2CO_3 (Li/K=62/38). This particular mixture of carbonate salts melts at 550°C, and when it infiltrates a porous lithium aluminate ($LiAlO_2$), a ceramic matrix, it functions both as an ion-conducting electrolyte and gasketing for the fuel cell stack. Negative carbonate ions (CO_3^{2-}) are responsible for conduction. As discussed above, the long-term performance is a cause for concern in MCFCs.

The following properties of the electrolyte have to be taken into account:

- Volatility of different alkali metal hydroxides generated in the moist cathode atmosphere
- Solubility of the cathode (NiO)
- Segregation of the ion species in the electrolyte
- Both oxygen and carbon dioxide solubility in the electrolyte
- Oxygen reduction kinetics

Loss of electrolyte mainly occurs at the cathode via hydrolysis

$$MeCO_3 + H_2O \longleftrightarrow 2MeOH + CO_2 \tag{17}$$

Li_2CO_3/K_2CO_3 electrolytes have been used without any significant developments or improvements since Ketelaar and Broers' work. The 62/38 eutectic composition has a relatively high volatility, causing the fuel cell to dry out. The partial pressure of MeOH varies with the square root of the water vapor to carbon dioxide vapor pressure ratio:

$$p(\mathrm{Me}_i\,\mathrm{OH}) = K_i(T)\sqrt{\frac{p_{\mathrm{H_2O}}}{p_{\mathrm{CO_2}}}}. \tag{18}$$

$K_i(T)$ is the equilibrium coefficient of the carbonate ion in the melt according to the equilibrium equation

$$CO_3^{2-} + H_2O \longleftrightarrow CO_2 + 2OH^- \tag{19}$$

The anode off-gas, on the other hand, can be mixed with air after combustion and be reused in the cathode chamber with a sufficiently high air excess. Here, oxygen and carbon dioxide are consumed in a molar ratio of 1:2 by the cathode process. The change in composition of the electrolyte due to segregation and volatility of certain species may result in a change in melting temperature. The electrolyte can solidify, sometimes allowing gases to break through, causing the fuel cell to malfunction. Carbonate ions from the electrolyte are used up in the reactions at the anode, so one has to compensate for their loss by injecting carbon dioxide at the cathode. Segregation occurs as the potassium concentration increases near the cathode. This leads to an increased cathode solubility and hence a decline in cell performance. Recent studies show that using Na instead of K can decrease the amount of segregation [5] in a molten binary salt with a common anion, where segregation occurs due to the difference in mobilities of the cations. The heavier and larger potassium and sodium cation is faster than lithium in mixtures that have a higher potassium mole fraction than $x(K_2CO_3) = 0.32$ (Chemla effect), whereas below this (isotactic) concentration, lithium is faster. The partial pressure of carbon dioxide has a much smaller effect on the current–voltage curves than does the partial pressure of oxygen. Mass transfer limitations are not observed for carbon dioxide, but they are for oxygen at low partial pressures. CO_2 transport also hinders the cathode operation much less than oxygen transport, since this gas has a higher solubility in carbonate melts than oxygen. Only the Li_2CO_3/Na_2CO_3 eutectic with much lower sodium vapor pressures can assure long-term performance of MCFCs. We can conclude that the Li/Na electrolyte is more reliable and safer than Li/K. Since the ionic conductivity of Li/Na carbonate melts is higher than that of Li/K carbonate melts, Li/Na is the preferred material for the electrolyte. MCFCs need a more complex design, compared to other technologies using a solid electrolyte, because their electrolyte is in the liquid phase.

External fuel processors are not needed for MCFCs, since the fuels can be reformed internally at the high operating temperatures. Internal reforming includes converting methane and steam into a hydrogen rich gas.

$$CH_4 + H_2O \longleftrightarrow CO + 3H_2 \tag{12}$$

At the anode, hydrogen reacts with the carbonate ions to produce water, electrons, and carbon dioxide:

$$H_2 + CO_3^{2-} \longleftrightarrow H_2O + CO_2 + 2e^- \tag{8}$$

The electrons are conducted away from the anode to the cathode by an external circuit to do useful work. Oxygen from the air and carbon dioxide from the anode react at the cathode with electrons to form water and carbonate ions:

$$\frac{1}{2}O_2 + CO_2 + 2e^- \longleftrightarrow CO_3^{2-} \tag{9}$$

The carbonate ions migrate through the electrolyte to the anode and complete the electrical circuit. CO_3^{2-} is used up at the anode, while CO_2 is needed at the cathode.

Electrode

A significant advantage of the MCFC is that non-noble metals can be used as electrodes. At the high operating temperature, a nickel anode and a nickel oxide cathode is able to promote the electrochemical reaction. This means lower production costs compared to low temperature fuel cells, where platinum has to be added to the electrode as a catalyst. The Ni electrodes are less prone to CO poisoning, hence coal-based fuels can be used, especially since internal reforming can take place. The main problem with the electrodes is their solubility in the electrolyte by Ostwald ripening, which is a dissolution/reprecipitation process. It decreases the internal surface area of the porous nickel oxide cathode, causing its performance to deteriorate. The solubility of nickel oxide (cathode material) is dependent on the cathode potential and temperature. The solubility of Ni and NiO in Li/Na carbonate melts was found to be lower than in Li/K carbonate melts. Although Li/Na carbonate melts have been found to have superior performance compared to Li/K carbonate melts, the lower accompanying oxygen solubility reduces the cathode performance on lean gas with oxygen partial pressure below 0.1 atm.

Outlook

Molten carbonate fuel cells are interesting candidates for stationary power generation networks, similar to the SOFC concept. The technology of the SOFC seems to be further developed and superior to MCFCs, shadowing the advantages that MCFCs can provide. Once SOFCs become ripe for commercialization, MCFCs will probably not be able to compete.

POLYMER ELECTROLYTE FUEL CELLS

The proton exchange membrane – also known as polymer electrolyte membrane (PEM) – fuel cell uses a polymeric electrolyte. The proton-conducting polymer forms the heart of each cell; electrodes, usually made of porous carbon with catalytic platinum incorporated into them, are bonded to either side of the electrolyte to form a one-piece membrane–electrode assembly (MEA). The following are some key advantages that make PEMs such a promising technology for the automotive market:

- Low temperature operation
- Quick start-up
- Absence of corrosive liquids

- Functions in any orientation (or zero gravity for that matter)
- Thin MEAs that allow compact cells

The PEM fuel cell was developed in the 1960s in the General Electric Research Laboratory. As with so many technologies, the space program and military-funded research fast-forwarded its development. PEM membranes were first applied to projects of the US Navy and the US Signal Corps. PEM cells were used in NASA's Gemini program, which was to serve as a means of testing the technology for the Apollo missions. Batteries were not suitable for a journey to the moon because of the extended flight duration. Early PEM systems were, however, unreliable and plagued with leakages and contamination. The systems installed in Gemini spaceships had an operational lifetime of just 500 h, although this was considered suitable. Another source of concern was the water management system, which is required to keep the membrane hydrated to the correct extent. Apollo designers opted for the more mature technology of AFCs, as did the space shuttle designers in the 1970s. Recently, however, as part of NASA's program of continuous upgrade on the Shuttles, PEM systems have replaced the aging AFC technology as the primary power source for the shuttles' systems. GE decided to abandon their research on PEMFCs in the 1970s, probably due to high cost. At that time, the catalysis required 280 g of platinum per m^2 of electrode, compared to the current figure of 2 g m^{-2}, or less.

Automobiles are arguably one of the most important consumer products on the planet. The finite fuel reserves, which they are chugging their way through, are not currently a limiting factor, but they will be soon. Much investment has been aimed at developing fuel cell technology for the automotive industry, and the electrolyte of choice is the PEM. We will look at the problems that automotive companies need to overcome before fuel cell cars become ripe for market introduction. The scope of PEMFCs is, arguably, wider than that of any other power supply technology, with the potential to power a range of devices from cell phones and laptops to busses, boats, and houses.

The PEMFC is constructed in layers of bipolar plates, electrodes, and membranes (Figure 1-9).

Each individual cell produces about 0.7 V EMF when operating in air. In order to generate a useful voltage, the electrodes of multiple cells must be linked in series. In addition to connecting the cells, we must ensure that reactant gases can still reach the electrodes and that the resistance of the electrodes has a minimal effect. A bipolar plate is used to interconnect the anode of one cell to the cathode of the next. It must evenly distribute reactant gases over the surface of the anode, and oxygen/air over the cathode. Bipolar plates may also need to carry a cooling fluid and, in addition,

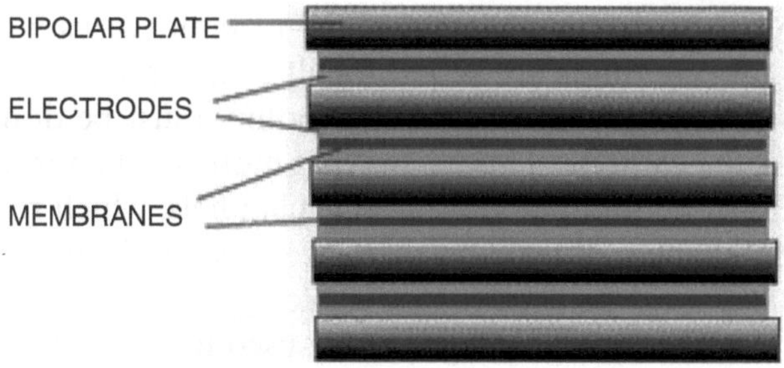

Figure 1-9. Membrane–electrode assembly

need to keep all these gases and cooling fluids separate. Design considerations are that

- The electrical contacts should be as large as possible
- The plate should be thin to minimize resistance
- Gas needs to flow easily across the plate

Often these factors are mutually competing, for example a large contact area would reduce the width of the gas channels.

In a simple plate design, as in Figure 1-10, the channels extend right to the edge, and the reactant gases are supplied to the system via external manifolding. Reactant gases flow at right angle to each other. External manifolding is a very simple solution and therefore very economic. But the technique has two major disadvantages: (1) the gaskets needed to seal the plates do not form a tight seal where the channels come to the edge of

Figure 1-10. Bipolar plate

the plate, leading to localized leaks of the reactant gases; (2) additional channels for cooling fluids are very difficult to incorporate into an externally manifolded system, so that all the cooling must be done by the air flowing across the cathode. This means that more air than is necessary for the reaction must be pumped through the channels, which requires wider channels. The chance of leaks is therefore increased, and some of the energy produced must be used to power blowers. While simplicity is usually a bonus, external manifolding is rarely used in modern systems. Most modern bipolar plates make use of internal manifolding. The three examples in Figure 1-11 show how this might be achieved. In each case, the channels do not run to the edge of the plates, allowing a gasket to be fitted and a gas-tight seal to be more easily achieved.

Referring to Figure 1-11:

- The design on the left is a fairly simple parallel channels design; reactant gases are blown into one end of the channel through one hole and removed at the other hole. There are many different designs possible, and designers of bipolar plates have yet to reach an agreement on which type is best. In parallel designs, water or gas may build up along one of the channels, causing a temporary blockage. In this case, the reactants will happily continue to pass through the other channels and not clear the blockage.
- The second design, a serpentine design, guarantees that if reactants are flowing at all, they are flowing all along the channel, and blockages are easily cleared. The problem in this case is that it takes more effort to push reactants through the long, torturous path.
- The third design is a compromise between the first two and is often encountered in bipolar plate designs. The channels are typically about 1 mm in width and depth. The pressure difference between the start and end of a channel must be engineered to overcome the surface tension of water droplets forming on the channel walls, in order to clear blockages.

Figure 1-11. Internal gasketing

The material properties of a bipolar plate, as summarized by Ruge and Büchi [6], must take into account several important factors:

- Electrical conductivity > 10 S cm^{-1}
- Heat conductivity of 20 W m^{-1} K^{-1} if the cooling fluid is integrated, 100 W m^{-1} K^{-1} if heat is removed only from the edges
- Gas permeability < 10^{-7} mbar L s^{-1} cm^{-2}
- Resistant to corrosion in an environment of acidic electrolyte, hydrogen, oxygen, heat, and humidity
- Reasonably high stiffness, $E > 25$ MPa
- As little cost as possible
- Thin dimensions for maximum stack volume
- Light weight for minimum stack mass
- Effective production process with a short cycle time

These difficult and sometimes conflicting specifications which must be met, along with the fact that modern electrodes require very little catalytic platinum, mean that the bipolar plate is the most expensive part of a modern fuel cell. In conclusion, we should note that although a broad range of bipolar plate manufacturing techniques exist, none of them fully meet the criteria set above. There is much development still to be done in this area before a new industry standard is established.

Electrolyte

Dupont's Nafion™ ion exchange membrane forms the basis of the PEMFC. Each company involved in the development of PEMFCs may have their own variant of Nafion; however, they are all based on the same sulfonated fluoropolymers, and Nafion remains the industry standard of comparison for membranes (although it is not always the most suitable). Nafion is a polymer based on polytetrafluoroethylene (PTFE, Figure 1-12), containing a fraction of pendant sulfonic acid groups ("sulfonic acid group" usually refers to the undissociated SO_3H group, whereas "sulfonate" refers to the ionized SO_3^- group after the proton has dissociated, see Figure 1-13).

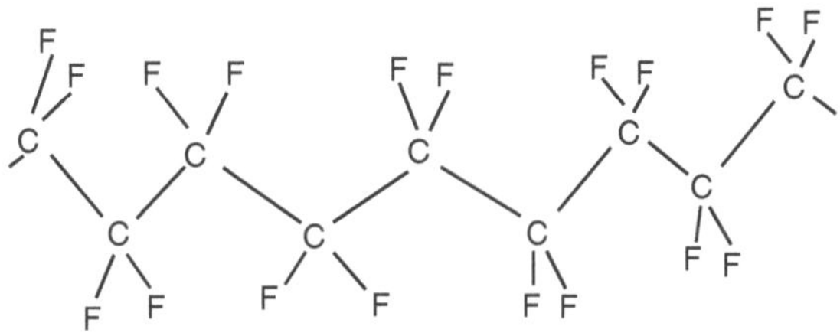

Figure 1-12. Polytetrafluoroethylene

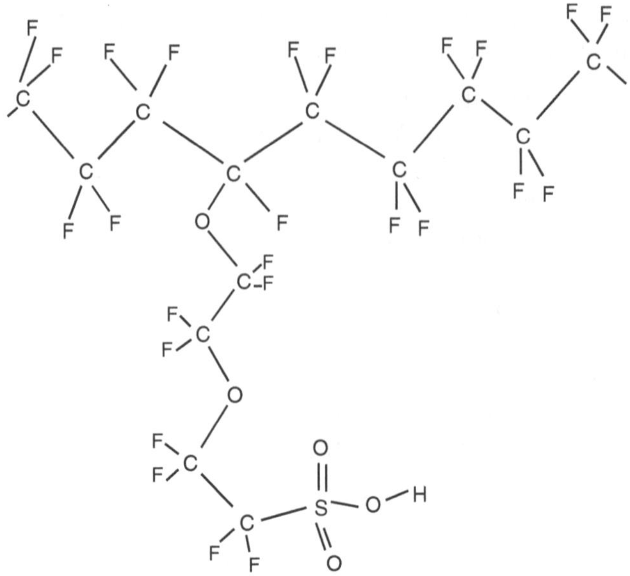

Figure 1-13. Nafion

The length of and the precise nature of the side chains vary between different brands of polymer. Common to all is the PTFE-based fluorocarbon "backbone" of the polymer that has several desirable properties:

- PTFE is hydrophobic – this means the hydrophilic sulfonate groups are effectively repelled by the chains and thus cluster together.
- PTFE is extremely resilient to chemical attack – important since the environment within the membrane is hostile and very acidic; hydrocarbon-based polymers would tend to degrade rapidly.
- PTFE is a thermoplastic with high mechanical strength – meaning very thin membranes can be produced, reducing the thickness of each cell and increasing the power density of the stack.

The mechanism of proton conduction through Nafion is easy to understand. The hydrophilic sulfonate groups of the polymer membrane cluster together and form lipophobic pockets. Water collects around the clusters of ionized sulfonate groups. Once the membrane is sufficiently hydrated, hydrogen ions dissociate and are free to migrate. The protons are actually strongly associated with water molecules and transported in the form of H_3O^+ hydronium ions, or even higher order cations. The Zundel ($H_5O_2^+$ – basically protonated water dimer) and the Eigen ($H_7O_3^+$) cations are thought to be particularly important in transfer of protons from one hydronium species to another.

Sulfonic acids are highly acidic (pK_a ≈ −6 in Nafion), i.e., they have a high tendency to dissociate into anions and protons. These protons act as the charge carriers through the membrane. The polymer must be hydrated to the correct degree in order to promote dissociation of ionic groups and provide a mechanism for proton transport. Proton conductivity is strongly dependent on the water content of the membrane. The water in the membrane is localized to the hydrophilic groups, where the protons dissociate, and is transported both in a vehicular manner (by diffusion of hydrated protons) and also structurally (via proton transfer between hydrated clusters). Typical PEMs have conductivity of order 0.01–0.1 S cm^{-1} at 80–90°C, which is a far lower temperature than other solid-state (usually ceramic) electrolytes operating temperatures.

Electrode

In the first fuel cells, platinum was used in relatively large quantities. This led to the mistaken belief that most of the cost of a fuel cell came from the platinum content. Generally this is no longer the case. Platinum particles are deposited very finely onto carbon powders, so that the platinum is very carefully dispersed with a maximal surface area. With catalysts produced in this way, the raw material platinum cost is just US $10 for a 1-kW cell stack. Before the catalyst layer is applied to the electrolyte, a coating of soluble electrolyte is brushed onto it. This ensures that there is good contact between the platinum and the electrolyte to achieve the important three-phase interaction between gas, catalyst, and electrolyte necessary for the reaction to proceed.

The catalyst (Figure 1-14) can be applied to the membrane in one of two ways: either directly onto the membrane in the form of a catalyst powder or alternatively by rolling, spraying, or printing, and adding the supporting electrode structure – often called the gas-diffusion layer – afterward, see Figure 1-15. The electrodes can also be assembled separately and bonded to the membrane in their complete form by hot pressing. The catalyst powder is sometimes mixed with PTFE to drive out product water and prevent the electrode from becoming water logged. A "gas diffusion layer" is added between the catalyst and the bipolar plate to provide some rigidity to the MEA and to ensure ease of diffusion. This layer is usually composed of carbon cloth or carbon paper 0.2–0.5-mm thick, with more PTFE added to expel water.

PEM is a solid-state electrolyte that functions at around 80°C. Compared to the 700–1,000°C at which the solid oxide electrolytes become conductive, this is a low temperature.

Figure 1-14. Catalyst made of carbon deposited with platinum particles

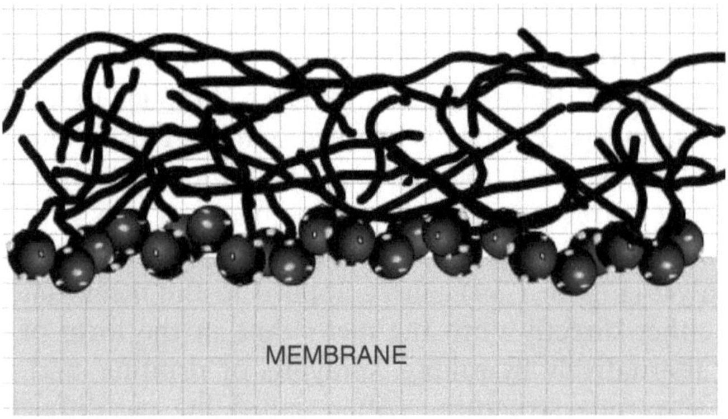

Figure 1-15. Membrane–electrode assembly powders

Outlook

PEMFCs are very popular among the car manufacturers. All fuel cell powered prototypes use this technology with Honda with its first commercially available fuel cells car pioneering the market.

THE IDEAL REVERSIBLE SOFC – BASIC DERIVATION OF POTENTIAL AND EFFICIENCY

The fuel cell is an electrical device, whose processes can be fully described by thermodynamic principles.

The first and second laws of thermodynamics describe the reversible SOFC. The reactants (fuel and air) supply a total enthalpy $\Sigma n_i H_i$. With total enthalpy $\Sigma n_j H_j$ leaving the fuel cell, the change in enthalpy is $\Delta H = \Sigma n_j H_j - \Sigma n_i H_i$. Heat q_{FC} has to be extracted from the fuel cell and the reversible work w_{FC} delivered (Figure 1-16).

$$\text{1st law of thermodynamics: } q_{\mathrm{FC}} + w_{\mathrm{FC}} = \Delta H \tag{20}$$

$$\text{2nd law of thermodynamics: } \int \mathrm{d}S = 0 \tag{21}$$

The reaction entropy has to be compensated by the transport of heat to the environment:

$$\mathbf{D}S - \frac{q_{\mathrm{FC}}}{T_{\mathrm{FC}}} = 0 \tag{22}$$

The reversible work from the above equations is

$$w_{\mathrm{FC}} - \Delta H - T_{\mathrm{FC}} \Delta S \tag{23}$$

The Gibbs enthalpy is equal to the reversible work of the reaction. The reversible efficiency is equal to the ratio of the Gibbs enthalpy to the reaction enthalpy

$$\eta_{\mathrm{FC}} = \frac{\mathbf{D}G}{\mathbf{D}H} = \frac{\mathbf{D}H - T_{\mathrm{FC}}\mathbf{D}S}{\mathbf{D}H} \tag{24}$$

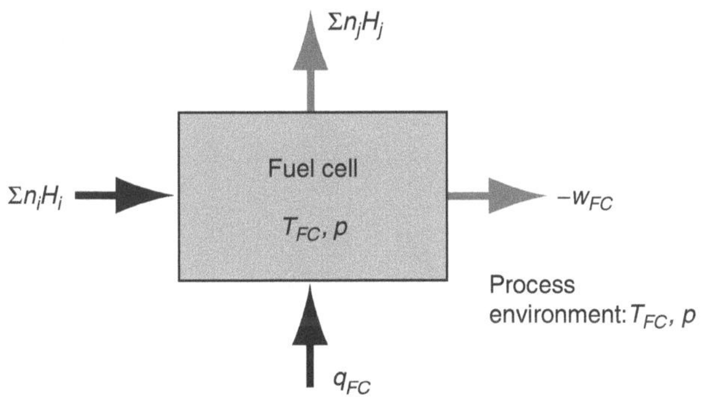

Figure 1-16. Thermal dynamics of SOFC

Hydrogen absorbed at the anode is ionized, and the electrons released are conducted away to do useful work. Oxygen atoms arriving at the cathode are ionized by the electrons coming from the anode. The protons and oxide ions react and form water:

Anode: $H_2 \longleftrightarrow 2H^+ + 2e^-$
Cathode: $\frac{1}{2}O_2 + 2e^- \longleftrightarrow O^{2-}$
The net reaction is: $2H^+ + O^{2-} \longleftrightarrow H_2O$

This shows that the molar flow of hydrogen is twice the molar flow of oxygen. The electric current is therefore

$$I = \dot{n}_{\text{el}}(-e)N_A = -2\dot{n}_{H_2}F \tag{25}$$

The electric current is a measure of the rate at which fuel is spent. The electric and thermodynamic quantities are matched by considering reversible power:

$$P = VI = \dot{n}_{H_2}w = \dot{n}_{H_2}\Delta G. \tag{26}$$

Hence the reversible voltage (Figure 1-17) is

$$V = \frac{-n_{H_2}\Delta G}{\dot{n}_{\text{el}}F} = \frac{-\Delta G}{\dot{n}_{\text{el}}F}. \tag{27}$$

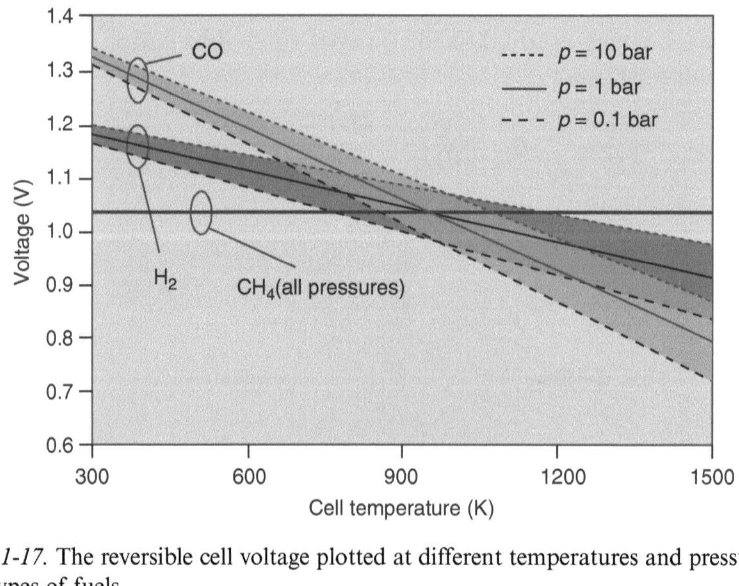

Figure 1-17. The reversible cell voltage plotted at different temperatures and pressures for three types of fuels

Using the assumption that we have (near) ideal gases, a more accurate equation for the Gibbs energy is

$$\Delta G(T,p) = \Delta H(T) - T\,\Delta S(T,p), \tag{28}$$

where

$$S(T,p) = S^0 + \int_{T_0}^{T} \frac{C_p(T)}{t}\,\mathrm{d}t - R\ln(p/p_0). \tag{29}$$

The resulting Gibbs energy is

$$\Delta G(T,p) = \Delta G(T) + TR\ln(K), \tag{30}$$

where K is the equilibrium constant. Similarly, the resulting Nernst potential is

$$V_\mathrm{N} = \frac{-\Delta G(T)}{n_\mathrm{el}F} - \frac{RT\ln(K)}{n_\mathrm{el}F}. \tag{31}$$

Chemical Thermodynamics of SOFCs

The reactant properties and fuel cell condition determines the electrical potential difference between the anode and the cathode electrode. The following analysis assumes quasistatic equilibrium and a steady flow of gas and ions.

At the cathode–electrolyte interface, the following half reaction occurs

$$\frac{1}{2}\mathrm{O_2} + 2\mathrm{e}^- \longleftrightarrow \mathrm{O}^{2-} \tag{7}$$

In equilibrium the sum of the chemical potentials is zero so that

$$\frac{1}{2}\mu_{\mathrm{O_2}} + 2(\mu_{e^-} - F\mathbf{U}_\mathrm{C}) = \mu_{\mathrm{O}^{2-}} - 2F\mathbf{U}_E. \tag{32}$$

Rearranging (32) gives

$$\mathbf{U}_\mathrm{C} - \mathbf{U}_\mathrm{E} = \frac{1}{2F}\left(\frac{1}{2}\mu_{e^-} - F\mathbf{U}_C\right) = \mu_{\mathrm{O}^{2-}} - 2F\mathbf{U}_\mathrm{E}. \tag{33}$$

At the anode–electrolyte interface, the half reaction

$$\mathrm{H_2} + \mathrm{O}^{2-} \longleftrightarrow \mathrm{H_2O} + 2\mathrm{e}^- \tag{6}$$

occurs. In equilibrium the sum of the chemical potentials is zero, so that

$$\mu_{H_2} + (\mu_{O^{2-}} - 2F\Phi_E) = \mu_{H_2O} + 2(\mu_{e-} - F\Phi_A). \tag{34}$$

Rearranging (34) yields

$$\mathbf{U}_A - \mathbf{U}_E = \frac{1}{2F}(\mu_{H_2O} + 2\mu_{e^-} - \mu_{H_2} - \mu_{O^{2-}}); \tag{35}$$

hence, the electric potential difference between the anode and the cathode is calculated as

$$\mathbf{U}_C - \mathbf{U}_A = \frac{1}{2F}\left(\mu_{H_2} + \frac{1}{2}\mu_{O_2} + \mu_{H_2O}\right). \tag{36}$$

Assuming that the gases behave like ideal gases, we can write for the chemical potential

$$\mu(T, p) = \mu^0(T) + RT\ln(p/p_0) + RT\ln(x_i), \tag{37}$$

where x_i is the mole fraction of component i and $\mu^o(T)$ is the chemical potential per mole of an ideal gas at $p_0 = 1$ bar.

Substituting (37) in (36), the potential difference in electric potentials becomes

$$\mathbf{U}_C - \mathbf{U}_A = \frac{1}{2F}\left(\mu^0_{H_2} + \frac{1}{2}\mu^0_{O_2} - \mu^0_{H_2O}\right) + \frac{RT}{2F}\ln\left(\frac{X_{H_2}\sqrt{X_{O_2}}}{X_{H_2O}}\right) + \frac{RT}{4F}\ln(p/p_0). \tag{38}$$

The typical value for the potential difference that is generated by the fuel cell is around 1 V. In order to obtain the higher voltages that are necessary to power everyday appliances, multiple individual cells must be connected in series, creating a fuel cell stack.

CONCLUSIONS

Fuel cell technology is at the doorstep of commercialization. The frequent introduction and celebration of fuel cell powered prototype cars, the demonstration of electronic gadgets, and the market introduction of stationary SOFC power generators for home use, all indicate the immense potential of and interest in fuel cells. Years of development and hard work are beginning to come to fruition, although some challenges still remain. It is always exciting to witness how an intellectual curiosity coming from academia revolutionizes an industry and its impact on the world.

ACKNOWLEDGMENTS

The author is indebted to Professor Yang Shao-Horn (MIT) for research supervision and for Professor Linn W. Hobbs (MIT), who critically read and edited the entire manuscript. The author gratefully acknowledges the financial support of the MIT Undergraduate Research Opportunities Program and the Cambridge-MIT-Exchange. A generous grant from the Cambridge-MIT-Institute and the Department of Materials Science and Metallurgy in Cambridge, UK, made possible the author's attendance at the IMAPS Fuel Cell Workshop.

References

[1] Renewables in Global Energy Supply, An IEA Fact Sheet, 2006.
[2] S.G. Meibuhr, Review of United States Fuel – Cell Patents Issued from 1860 to 1947, Res. Labs., Gen. Motors Corp., Warren, MI, Electrochimica Acta, 1966.
[3] N.Q. Minh, Centenary of Nernst's Discovery of Zirconia Electrolytes – Review of Zirconia-Based Electrochemical Technologies, Allied Signal Aerospace, Torrance, CA, USA. *Proceedings – Electrochemical Society*, 1999.
[4] H.-H. Mobius, On the history of solid electrolyte fuel cells. *Journal of Solid State Electrochemistry*, 1 (1997) 2–16.
[5] H. Wendt, T. Brenscheidt, and M. Kah, Different molten alkali carbonate eutectics as fuel cell electrolytes for MCFCs, *Molten Salts Bulletin No.*, 67 (1998) 2–17.
[6] M. Ruge and F.N. Büchi, PE fuel cells: evaluation of concepts for a bipolar plate design and construction, *Proceedings – Electrochemical Society*, 2001, pp. 165–173.

Bibliography

Nathan C. Lewis, Global Energy Perspective, presentation given at California Institute of Technology, May 2005, available at http://nsl.caltech.edu/energy.html
Claude Mandil, The Energy Mix of a Sustainable Energy Future, presentation given at the *Delhi Sustainable Energy Summit*, New Delhi, India in 2–4 February 2006, available at http://www.iea.org/textbase/speech/2006/Mandil/DSDS.pdf
IEA, Renewables in Global Energy Supply – An IEA Fact Sheet, 2004, available at http://www.iea.org/textbase/papers/2006/renewable_factsheet.pdf
IEA Advanced Fuel Cells – Implementing Agreement Annual Report 2004, April 2005, available at http://www.ieafuelcell.com/html/publications.htm
J. Larminie and A. Dicks, *Fuel Cell Systems Explained*, John Wiley & Sons Ltd., Chippenham, UK, 2003.

S.C. Singhal and K. Kendall, *High Temperature Solid Oxide Fuel Cells: Fundamentals, Design and Applications*, Elsevier, Bodmin, UK, 2003.

T.D. Kaun, I.D. Bloom, and M. Krumpelt, Nonsegregating Electrolytes for Molten Carbonate Fuel Cells, Proceedings of the '97 Fuel Cell Review Meeting, Morgantown, West Virginia, August 26–28, 1997, available at http://www.netl.doe.gov/publications/proceedings/97/97fc/FC7-3.PDF

H. Wendt, T. Brenscheidt, and M. Kah, Different Molten Alkali Carbonate Eutectics as Fuel Cell Electrolytes for MCFCs, *Molten Salts Bulletin* No. 67 (1998) 2–17.

M.S. Dresselhaus et. al. Nature insight: materials for clean energy, *Nature* 414 (2001) 331–352.

K. Denbigh, *The Principles of Chemical Equilibrium*, 4th edn., Cambridge University Press, 2002.

A.J. Bard and L.R. Faulkner, *Electrochemical Methods – Fundamentals and Applications*, 2nd edn, John Wiley & Sons Ltd. USA, 2000.

Chapter 2

STABLE GLASS SEALS FOR INTERMEDIATE TEMPERATURE (IT) SOFC APPLICATIONS

Qingshan Zhu, Lian Peng, and Tao Zhang
Multiphase Reaction Laboratory, Institute of Process Engineering, Chinese Academy of Sciences, Beijing 100080, China

INTRODUCTION

Solid oxide fuel cells (SOFCs), which convert chemical energy of incoming fuels to electricity via electrochemical reactions, have become increasingly attractive to the utility and automotive industries for a number of reasons, including their high efficiency and low emissions [1]. Among different types of SOFCs, planar SOFC designs have several advantages over tubular designs [2], such as easier to be manufactured, lower processing cost, shorter current paths, and consequently higher power densities, etc. Planar SOFCs have therefore been extensively investigated over the past decade. In the planar design, a gas-tight seal is required to keep the fuel gas and air separated from each other to prevent direct combustion which decreases power generation efficiency and causes local overheating. The seals must be stable in both oxidized and reduced atmospheres, be chemically compatible with other fuel cell components, and have matched thermal expansion coefficients (TEC) with other components. Moreover, these properties must be kept unchanged over the required lifetime of at least 40,000 h at elevated temperatures, together with hundreds of thermal cycles. Although many advances have been achieved for SOFC sealing, none of the current sealing materials can meet all the above criteria at the same time. Accordingly, sealing has been identified as one of the toughest technical challenges for the development of planar SOFCs [3].

In the present chapter, we will begin with a brief review of the state-of-the-art sealing concepts and sealing materials, followed by identifying the major problems for glass-based sealing materials. Then our new approach for sealing glass development will be introduced. Finally, the TEC, chemical compatibility with 8YSZ (8 mol% yttria stabilized zirconia) and thermal stability of the newly developed sealing glasses will be discussed in detail.

STATE-OF-THE-ART SOFC SEALING

Over the past decade, different SOFC sealing concepts have been proposed. Among these sealing concepts, rigid bonded sealing and compressive sealing have been extensively investigated [4], together with the development of suitable materials.

Rigid Sealing

Rigid sealing refers to the concept that SOFC components are rigidly bonded together by sealants, where after sealing the bonded components cannot move against each other. Two types of materials, glass-based and alloy-based (braze) materials, have been investigated for this sealing concept. The status quo of these two types of sealing materials will be summarized below.

Glass/Glass–Ceramic Sealants

Glass or glass–ceramic materials are the most commonly used sealants for SOFCs because the properties can be easily tailored by manipulating glass compositions, and also because glasses are cost effective and simple to be processed. Glasses are typically distinguished by the glass formers, like phosphate glasses, borate glasses, and silicate glasses. All the three types of glasses have been explored for SOFC sealing. It has been generally showed that at high temperatures, glasses purely based on phosphate are easily volatilized and tend to react with the Ni-YSZ anode to form nickel phosphate and zirconiumoxyphosphate [5,6]. Also, phosphate glasses usually crystallize to form meta- or pyrophosphates, which exhibit low stability in a humidified fuel gas at SOFC operating temperatures. Consequently, glasses purely based on phosphate are not suitable for SOFC sealing applications [7]. Borate-based glasses and glass–ceramics have also been investigated as SOFC sealing materials. The problem for glasses with boron oxide as the only glass former lies in the fact that borate tends to react with a humidified fuel atmosphere to form volatile species like $B_2(OH)_2$ and $B_2(OH)_3$ at SOFC operating temperatures, which causes loss of the glass seal [8]. It has been shown that a borate based glass exhibited a weight loss up to 20 wt% in a wet fuel atmosphere [5] and extensive interactions with other cell components both in air and wet fuel gas. Silicate glasses are most attractive as potential sealing materials because these glasses have much better chemical compatibility and stability as compared with phosphate- and borate-based glasses. Most sealing glass R&Ds were therefore focused on silicate glasses and glass–ceramics. Some typical glass and glass–ceramic seals reported in literature [7–13] are summarized in Table 2-1.

Among various properties, glass transition temperature (T_g) and TEC are most extensively investigated, as these two properties directly determined sealability of a glass. T_g is the temperature where the transition of a material from the solid state into the liquid state takes place. T_g is often determined via dilatometry and roughly corresponds to a viscosity of ~10^{12} Pa s, below which the material is "soft" enough to release thermal stress through plastic deformation. So, the T_g is the indicator of the "softness" of a glass. For SOFC applications, it is desirable that the T_g be as low as possible since significant thermal stress only starts to build up below the T_g. On the other hand, the sealant viscosity at fuel cell operating temperatures must be greater than 10^3 Pa s, below which the sealant would flow readily, to provide sufficient sealing strength [14]. So, a suitable sealing glass and glass–ceramic should flow sufficiently to provide an

Table 2-1. Compositions of glass/glass–ceramic sealants (mol%)

Glass former	Al_2O_3	Others	T_g (°C)	TEC (10^{-6} K^{-1})	Ref.
$52P_2O_5$	16	32MgO	–	–	[7]
$5SiO_2 – 50P_2O_5$	15	30MgO	–	–	
$18SiO_2 – 43P_2O_5$	13	26MgO	–	–	
$30SiO_2 – 37P_2O_5$	11	22MgO	–	–	
$8.11SiO_2 – 40.29B_2O_3$	6.92	$24.6SrO–20.1La_2O_3$	740–780	11.5	[8]
$5.72SiO_2 – 46.04B_2O_3$	17.69	$26.8SrO–3.75La_2O_3$	560	9.69	
$11.23SiO_2 – 27.01B_2O_3$	16.27	$41.76SrO–3.74La_2O_3$	700	9.2	
$28.29SiO_2 – 18.5B_2O_3$	14.26	$35.31SrO–3.65La_2O_3$	740–750	8.08	
$34.4SiO_2 – 17.2B_2O_3$	10.3	$36.1BaO–2La_2O_3$	627	10.1	[9]
$33.3SiO_2 – 16.7B_2O_3$	10	$35BaO–5La_2O_3$	656	11.1	
$33.3SiO_2 – 16.7B_2O_3$	10	$35BaO–5ZrO_2$	655	9.3	
$30.1SiO_2 – 21.5B_2O_3$	10.3	$36.1BaO–2ZrO_2$	614	10.6	
$34.4SiO_2 – 17.2B_2O_3$	10.3	36.1BaO–2NiO	620	11.5	
$33.3SiO_2 – 16.7B_2O_3$	10	35BaO–5NiO	615	12.8	
$50.5P_2O_5 – 11.3B_2O_3$	12.6	25.6MgO	603	5.9	[10]
$50.5P_2O_5 – 11.3B_2O_3$	12.6	$25.6CaO–5.0Cr_2O_3$	605	7.9	
$46.5P_2O_5 – 11.3B_2O_3$	11.6	25.6MgO+5.0CaO	618	5.7	
$42.5P_2O_5 – 11.3B_2O_3$	10.6	$25.6MgO–5.0Cr_2O_3$	620	5.9	
$21.45SiO_2 – 6.66B_2O_3$	5.39	56.1BaO–7.1CaO	–	–	[11]
$50SiO_2 – 14B_2O_3$	11	$24BaO–0.5As_2O_3$	–	–	[12]
$47.5SiO_2 – 13.3B_2O_3$	10.4	$22.8BaO–5MgO–0.5As_2O_3$			
$45SiO_2 – 22.6B_2O_3$	9.8	$21.6BaO–10MgO–0.5As_2O_3$	–	–	
$60SiO_2$		32BaO–8MgO	703	9.8	[13]
$60SiO_2$		32BaO–8ZnO	670	10.5	
$50SiO_2$		40BaO–10MgO	686	12	
$50SiO_2$		40BaO–10ZnO	676	11.5	

adequate seal and to release the thermal stress as much as possible, while maintaining sufficient rigidity to achieve mechanical integrity. T_g of a glass would decrease with increasing the contents of boron oxide, alkali metal oxides as well as alkaline earth metal oxides.

The TEC is another important criterion for the selection of sealing glasses and glass–ceramics as TEC mismatch is the origin of thermal stress. TEC values are in the range of $10.5–12.5 \times 10^{-6}\ K^{-1}$ (room temperature to 800°C) for SOFC components (anodes, 8YSZ electrolyte, cathodes, interconnects), so it is generally required that the TEC of sealants should be also in the range of $10.5–12.5 \times 10^{-6}\ K^{-1}$. Normal silicate glasses usually have much lower TEC values and it is not easy to obtain a glass of high TEC values together with suitable T_g. The TEC of silicate glasses will normally increase with increasing alkali metal oxides contents. However, Li^+, Na^+, and K^+ cations are easily diffused into other fuel cell components, which cause cell performance degradation. Consequently, alkali metal oxides addition is not recommended in most cases [14]. To achieve high TEC values, alkaline earth metal oxides are often used [9], among which BaO addition is most attractive due to its relatively large contribution to the TEC increase among all alkaline earth metal oxides. Another way to increase sealants' TEC is through deliberate crystallization control of a glass to form a glass–ceramic. Glasses are not thermodynamically stable and tend to crystallize at SOFC operating temperatures. However, uncontrolled crystallization usually results in unwanted phases that might have too low or too high TECs. The crystallization is a thermodynamically favorable and kinetically controlled process. To obtain suitable crystalline phases, it often needs to add nucleating agents to induce the nucleation of certain phases. Also, crystallization kinetics, like nucleating temperature and time, crystallization temperature and time, is critical to the property control. It has been shown that the formation of Ca_2SiO_4, $BaSiO_3$, $BaSi_2O_5$, and $Ba_3CaSi_2O_8$ phases would increase the TEC as these phases have high TECs, while phases like hexacelsian ($BaAl_2Si_2O_8$), celsian, and cordierite ($Mg_2Al_4Si_5O_{18}$) can significantly decrease the TEC [4].

Glass-based sealants must be chemically stable in both air and fuel gas atmospheres throughout the service time of SOFC. For the phosphate glasses and borate-based glasses, it has been shown that the stability is insufficient even for a short time of exposure, especially in wet fuel atmospheres. However, chemical stability for silicate glasses is seldom reported and needs to be further investigated. The sealants must be chemically compatible with other fuel cell components. Investigations showed that the chemical compatibility of glass–ceramic sealants with the 8YSZ electrolyte was generally good, while severe reactions were commonly observed between glass-based sealants and metal interconnects. The sealant–interconnect interaction is problematic because such reactions cause changes in both glass phase and the interconnect phase around the interface. It has been demonstrated that small composition variation of glasses or metals could result in significant difference in interface composition and structure [15]. For example, investigations revealed that small amounts (<0.5 wt%) of Al, Si, Mn, and Ti in the interconnect metals have great influence on the interaction rate. It had also been shown that pores up to several hundred micrometers were observed after only a few hours contact of SUS 430 with a sealing glass at 750–800°C [11]. The interaction mechanism among different metals and glasses has not yet been fully understood and needs further efforts to get a profound understanding about these interactions.

Thermal stability is another very important property for glass-based sealants since seals need to function for at least 40,000 h. As mentioned previously, glasses are thermodynamically unstable and have the tendency to transform to thermodynamically stable phases through crystallization during exposure at SOFC operating temperatures. The transformation is accompanied with thermal property changes, which might enlarge the TEC mismatch between the sealant and the materials being sealed. For example, the TEC value decreased up to 36% for a glass after being heat treated at 800°C for 1,000 h [9].

Brazes

The other approach to a rigid seal is a metallic braze. Metallic materials have lower stiffness as compared to ceramics and can undergo plastic deformation more easily, both of which allow for accommodation of thermal and mechanical stresses. Brazes are often based on alloys of Pt, Au, Ag, Pd, Cu, Ni, Ti, and Cr. Of these, precious metal-based alloys are most promising for use in SOFCs, as these brazes are capable of withstanding fuel and oxidizing atmospheres at high temperatures [16]. One of the major problems in obtaining a good metal-ceramic sealing is inadequate wetting of the ceramic by braze metals. Approaches employed to solve the problem include the addition of reactive elements like Ti, Hf, Zr, etc. and the addition of metal oxides such as CuO, Al_2O_3, Al_2TiO_5, etc. [17]. The addition of metal oxides could also be used as a mean to adjust the TEC of brazes, as the TECs of most brazes are in the range of 16–21 $\times 10^{-6}$ K^{-1}, which are much higher than those of 8YSZ and other SOFC components. However, brazes are electrically conductive, so proper insulation should be achieved to avoid short-circuit when applied to SOFCs. Brazes also have insufficient long-term oxidation resistance at high temperature under SOFC atmospheres.

Compressive Seals

Another possible alternative to glass-based seals is the use of compressive, nonbonding seals. If the seals are nonbonding, the individual stack components are free to expand and contract during thermal cycling. Compressive seals utilize materials such as sheet-structure silicates, which do not bond to the SOFC components; instead, hermeticity is achieved by applying a compressive force to the stack, using suitable sealing materials as the gasket. The use of compressive seals brings several new challenges to SOFC stack design, like a load frame must be included to maintain the desired level of compressive load during operation, and the stack components must be able to withstand the compressive load required for adequate sealing for the lifetime of the stack. Compressive seal materials are mostly two types of materials: metallic materials and mica-based materials [18]. As mentioned before, metallic materials are seldom reported due to their electric conductivity. The research in this area was mainly focused on mica-based materials.

Micas belong to a class of minerals known as phyllosilicates, and are composed of sheets of silicate tetrahedrons. They are generally known for their high resistivity and uniform dielectric constant/capacitance stability, and consequently are used extensively in electronic devices [19]. Micas can accommodate large thermal stress because they do not bond to the SOFC components. However, mica seals alone normally exhibit high leak rates several orders of magnitudes than those of glass-based seals [20]. It is therefore necessary to combine with other materials to reduce the leak rate to an acceptable level for SOFC applications. Silver and glass were commonly used as the composite materials to improve the hermeticity of mica seals. Figure 2-1 shows the effect of various interlayers on the leak rate (sccm cm^{-1}, i.e., standard cubic centimeter per minute per unit leak length of seal) [21]. Clearly, the use of a glass as the interlayer for hybrid mica seals is more effective than silver interlayer in blocking or filling the surface defects such as grooves and voids presented at these interfaces. Moreover, leak rate decreases with increasing compressive stress.

Although the combination of glass with mica was found to be attractive, mica seals were also faced on many challenges. In general, seals with elastic or plastic properties require a permanent compression load to achieve gas tightness. In most cases, the sealing load is higher than that required for electrical contact between electrodes and interconnects, so an adaptation of the stack design to the requirements of a compressive seal is a prerequisite for application of this concept.

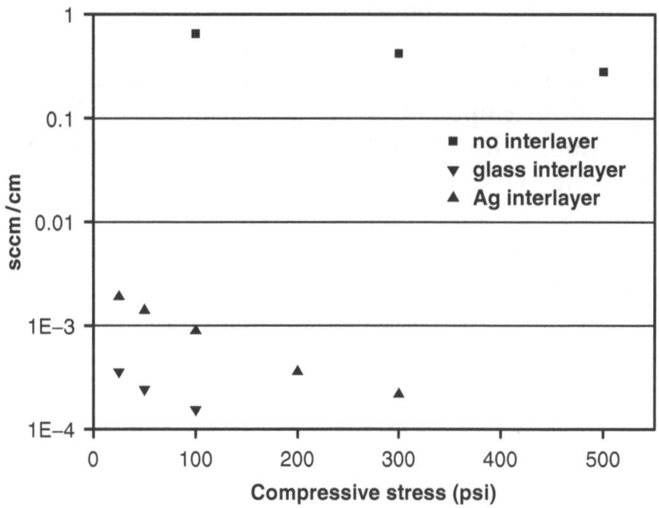

Figure 2-1. Effect of the applied compressive stress on the normalized leak rate of muscovite single-crystal mica with and without interlayer at 800°C

Challenges for Glass-Based Sealants

Due to the extremely stringent requirements for the sealing material, e.g., seal materials should be stable and capable of a service life of more than 40,000 h and hundreds of thermal cycles for stationary systems, or at least 5,000 h and 3,000 thermal cycles for transportation systems, the fundamental difficulty in fabricating planar SOFCs is how to effectively seal the anode/electrolyte/cathode assembly together with the interconnect to create a hermetic and stable stack. Glass or glass–ceramic materials are the most commonly used sealants for SOFCs. Although many advances have been achieved for glass-based seals, several challenges remain unsolved. Firstly, glass-based seals have very poor thermal stability, where the thermal properties of glass based seals change continuously with the time of exposure at high temperatures. As previously mentioned, glasses are easily crystallized at SOFC operating temperatures, which might cause a significant change about the thermal expansion coefficient. As demonstrated by Sohn et al. [9], the TEC change can be more than 36% after being kept at 800°C for 1,000 h. The change of the TEC value and other thermal properties can induce significant thermal stress at the sealing interface, and will ultimately cause the failure of the sealing as well as the stack when the thermal stress reaches a critical point. So, the thermal stability of current state of the art sealing glass/glass–ceramics needs to be further improved. Secondly, glass-based seals often have poor chemical compatibility with other SOFC components (especially with interconnects) at operation temperatures. As demonstrated by Yang et al. [11], extensive reactions with $BaCrO_4$ as the main resultant product were observed between a glass-based seal and metal alloys, where the reactions generated pores up to 200 μm aligned along the metal–glass interface. In some cases, the reaction also caused separation of the glass with the alloy matrix, possibly due to the thermal expansion mismatch between $BaCrO_4$ and metal interconnect. The extensive formation of interfacial pores and interface separation will greatly reduce interfacial bonding strength, which would be detrimental to structural stability of the SOFC stacks. Therefore, glass-based seals with much better thermal stability and chemical compatibility are still needed to be developed.

APPROACH FOR SEALING GLASS DEVELOPMENT

Conventionally, glasses were developed using the "trial and error" approach. However, such an approach does not work effectively for sealing glass development as sealing glasses usually have quite complex compositions and have to meet several requirements, such as TEC match, suitable viscosity, chemical compatibility with other materials, chemical stability under fuel cell atmospheres, long-term thermal stability, etc. For such complex composition and multiobjective glass development, it is already difficult enough to find a composition that could meet all the requirements, and it is even more difficult to optimize the composition to fit each requirement using the "trial and error" approach, which might be the primary reason for the slow advancement of glass-based seals over the past decade. A new approach, capable of quantitative design of sealing glasses, would be beneficial to speeding up the sealing glass development and to accelerating the commercialization of SOFCs as well.

In the present study, a new approach, aiming at quasiquantitative design of the chemical compatibility, chemical stability, TEC, and viscosity for sealing glasses, has therefore been proposed. As illustrated in Figure 2-2, a combined model, coupled from the TEC model and the viscosity model [22], will be first established. A glass composition could then be designed to meet the TEC and viscosity requirements. In the meantime, a thermodynamic model will also be constructed for predicting various reactions between a glass and other materials (anode, electrolyte, cathode, etc.). The glass composition from the combined model would then be used as the input for the thermodynamic investigation to check the chemical

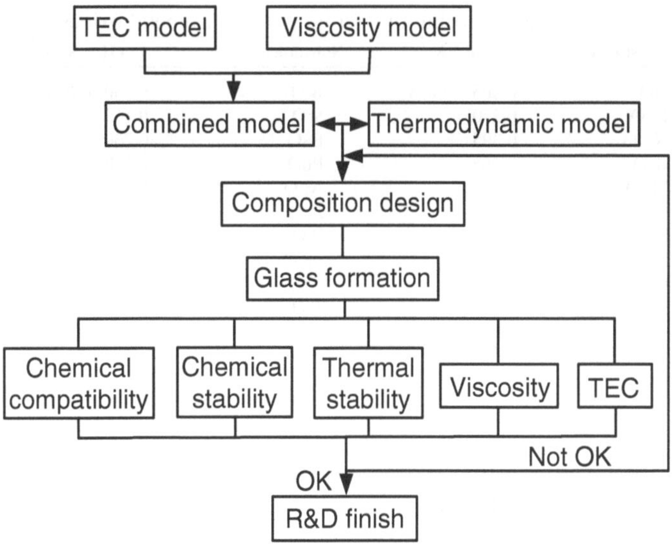

Figure 2-2. A schematic procedure for sealing glass development

compatibility between the glass and other materials, and the composition would be adjusted according to the results from thermodynamic analyses. The composition would then be redesigned using the combined model, taking the composition from the thermodynamic analyses as the constraint. Through the simulation between the combined model and the thermodynamic model, a composition that could meet all the requirements may hopefully be obtained. A glass will be made according to the composition, and its properties, including chemical compatibility with other materials, thermal stability, viscosity and TEC, etc., will be experimentally checked to validate the composition design. If any of the properties failed to meet the requirements, the glass design would start from the beginning again, taking into account the information obtained from the validation experiments.

The TEC of a glass is normally predicted by the weighted model as expressed by (1):

$$\alpha = \sum_i \alpha_i c_i, \tag{1}$$

where α is the TEC of the glass (10^{-6} K^{-1}), α_i is the TEC contribution factor of the *i*th component (10^{-6} K^{-1}) and c_i is the mole fraction of the *i*th component in the glass. Some TEC contribution values of common oxides are listed in Table 2-2. The viscosity of glasses against temperature and

Table 2-2. TEC contribution factors for various metal oxides [22]

Substance	α_i (10^{-6} K^{-1})	Substance	α_i (10^{-6} K^{-1})
SiO_2	0.005 – 0.038	B_2O_3	–0.050 to 0.00
Li_2O	0.270	SrO	0.160
Na_2O	0.395	BaO	0.200
K_2O	0.465	PbO	0.130 to 0.190
MgO	0.060	Al_2O_3	–0.030
CaO	0.130	ZrO_2	–0.060
ZnO	0.050	P_2O_5	0.140
MnO	0.105	Fe_2O_3	0.055
CoO	0.050	NiO	0.050
SnO_2	–0.045	TiO_2	–0.015 to 0.03
CuO	0.030	Ga_2O_3	–0.020

composition is normally expressed by the so-called Vogel–Fulcher–Tammann (VFT) equation as:

$$\log \eta = -A + B/(T - T_0). \tag{2}$$

Investigations by Lakatos et al. [22,23] demonstrated that the constants A, B, and T_0 can be well represented by simply additive terms for a certain range of compositions close to those used for containers and flat glasses as:

$$A = 1.713 + \Sigma a_i c_i, \tag{3}$$

$$B = 6237.01 + \Sigma b_i c_i, \tag{4}$$

$$T_0 = 149.4 + \Sigma t_i c_i, \tag{5}$$

where a_i, b_i, t_i are the VFT coefficients for the ith component and c_i is the concentration of the ith components in a glass.

Our investigations showed that the viscosity model and the TEC model could be easily combined together using a weighted sum strategy after normalization of the TEC data and the viscosity data, and it was quite easy to obtain a glass composition that meets the targeted values for both the viscosity and TEC models using an optimization algorithm [24]. The problem for the glass composition design using the combined model however lies in the fact that these two models are only valid for a certain range of compositions and the prediction accuracy is thus questionable for compositions out of the range. Investigations revealed that the prediction accuracy for these two models was actually very poor for glasses relevant to SOFC seals. For example, for the glass of $30.1SiO_2$–$21.5B_2O_3$–36.1BaO–$10.3Al_2O_3$– $5.0ZrO_2$ composition (mol%), the TEC model predicts a TEC value of 7.0×10^{-6} K^{-1}, whereas the TEC was measured to be

10.6×10^{-6} K^{-1} [9] (room temperature to 700°C), indicating that the coefficients in Table 2-2 are not completely applicable to this composition. It is therefore necessary to amend the coefficients for both the viscosity model and the TEC model in order to make them applicable to those compositions relevant to SOFC glass seals.

The chemical compatibility of a glass with other materials and chemical stability of a glass can be investigated by thermodynamic investigations. While $\Delta G < 0$ could be used to judge whether a reaction could proceed for simple systems, reactions occurred in complex systems like sealing glass would normally be simulated using minimizing the free energy method [25] as indicated by (6) and (7).

$$\text{Min: } G = \sum_{j=1}^{M} \sum_{i=1}^{N_j} n_i \mu_i, \tag{6}$$

$$\text{Conservation of Mass: } \sum_{j=1}^{M} \sum_{i=1}^{N_j} a_{ik} n_i = B_k \quad (k = 1, 2, \ldots, NN), \tag{7}$$

where G is the total free energy of the system, M is the total phases in the system, n_i is the molar quantity of the ith substance in a specified phase, N_j is the total number of substances in the jth phase, μ_i is the chemical potential of the ith substance under the specified condition, NN is the number of elements in the system, a_{ik} is the molar quantity of the kth element in the ith substance in a specified phase and B_k is the total molar quantity of the kth element in the system. The μ_i is expressed differently for gas phase, liquid phase and pure condense phase by (8), (9) and (10), respectively.

$$\mu_i = \mu_i^0 + RT \ln f_i, \tag{8}$$

$$\mu_i = \mu_i^0 + RT \ln a_i, \tag{9}$$

$$\mu_i = \mu_i^0, \tag{10}$$

where μ_i^0 is the chemical potential of the ith substance under the standard state, f_i is the fugacity of the ith gas species and is equal to partial pressure p_i for an ideal gas, a_i is the activity of ith liquid species and is equivalent to the mole fraction x_i for an ideal solution, R is the universal gas constant, and T is gas temperature in Kelvin.

The accuracy of thermodynamic simulation is dependent on thermodynamic model. It is quite easy to model the gas phase and the pure condense phases, while the liquid solution of oxide glasses have been difficult to model because of strong interactions between constituents, and it has been found that simple solution models do not accurately reproduce the thermodynamic and phase relations in most oxide liquid systems [26]. In the past two decades two approaches, the modified quasichemical model by Pelton et al. [27,28] and the modified associate species model by Spear et al. [29], were proposed for the simulation of oxide melts. The modified associate species model is attractive because it has been demonstrated that it can accurately represent the thermodynamic behavior of some complex chemical systems over wide temperature and composition ranges [26]. The basis for the modified associate species approach is that the complex oxide solutions can be represented by an ideal solution of end-member species and intermediate associate species. The intermediate associate species can be fictive species whose thermodynamic data can be determined by a reconstruction of an experimentally determined phase diagram through an optimization process.

To simulate a glass system relevant to our sealing glass development, the SiO_2–B_2O_3–BaO–CaO–Li_2O–ZrO_2 system was tentatively modeled using the associate species approach, where the model consists of 22 gas species including O_2, N_2, NO, NO_2, B_2O(g), B_2O_2(g), AlO(g), Al_2O(g), AlO_2(g), Al_2O_2(g), B_2O_3(g), CaO(g), SiO(g), $AlBO_2$(g), ZrO_2(g), ZrO(g), ZrN(g), La_2O_2(g), La_2O(g), BaO(g), Ba_2O(g), Ba_2O_2(g); and 19 pure solid phases including B_2O_3(cr), CaO(cr), Al_2O_3(cr), SiO_2(cr), CaB_4O_7(cr), Al_2SiO_5(cr), $CaAl_2SiO_6$(cr), ZrO_2(cr), $ZrSiO_4$(cr), $CaZrO_3$(cr), ZrN(cr), ZrB_2(cr), $BaZrO_3$(cr), Ba_2ZrO_4(cr), $Y_2Zr_2O_7$(cr), La_2O_3(cr), Y_2O_3(cr), BaO(cr), BaO_2(cr). For the liquid phase, apart from the end member species of B_2O_3(l), CaO(l), Al_2O_3(l), SiO_2(l), Li_2O (l), ZrO_2(l), CaO(l), BaO(l), seven associate species of $CaSiO_3$(l), $Al_4B_2O_9$(l), CaB_4O_7(l), Ba_2ZrO_4(l), $Al_2Si_2O_7$(l), $Li_2B_4O_7$(l), $LiBO_2$(l), $LiAlSi_2O_6$(l) were included. Thermodynamic data for the end members and the associate species were taken from references [30,31], no attempts were made to evaluate more associate species according to various phase-diagram at this stage. Gibbs free energy of Ba_2ZrO_4(cr) were estimated by the regular solution model. The thermodynamic model was used to predict chemical compatibility of various glasses with 8YSZ at high temperatures (at least above the liquid temperature of glasses) during the glass development.

EXPERIMENTAL

Glass Preparation

Glasses were prepared from reagent-grade Al_2O_3, SiO_2, B_2O_3, La_2O_3, ZrO_2, Y_2O_3, $BaCO_3$, $CaCO_3$, and $MgCO_3$. The oxides and carbonates in prescribed composition were melted in a Pt crucible at 1,250–1,400°C in air for 2–4 h, followed by quenching the glass melt in water. The as-quenched glass was remelted at 900–1,000°C for 0.5 h and cooled to room temperature naturally to obtain a glass block. The glass block was then cut and ground into bars with nominal dimensions of 15 × 4 × 2 mm^3 for various characterizations.

Heat Treatment

Heat treatments were performed for the as-prepared glasses to obtain glass–ceramics. Nucleating agents, including ZrO_2, TiO_2, Cr_2O_3, P_2O_5 were added up to 5 wt% to their parent glasses through comelting with the parent glasses. The heat treatment procedure was determined based on glass transition temperature (T_g), crystallization temperature (T_c) and liquid temperature (T_{liq}) of glasses, which were determined by differential thermal analyses (DSC-TG, Netzsch STA 449C, Germany). In a typical experiment, after melting a glass together with a proper amount of a nucleating agent, the glass was first heated at a heating rate of 10°C min^{-1} to ~690°C for 0.5–2 h to achieve homogenous initiation of crystallization, then to a temperature between T_c and T_{liq} for 2–10 h to obtain a glass–ceramic, followed by cooling down to room temperature at a cooling rate of 10°C min^{-1}. The as heat-treated glass was then subjected to various characterizations to reveal the nature of the as-obtained crystals.

Glasses and Glass–Ceramics Characterization

All the as-prepared glasses were characterized by X-ray diffractometry (XRD, X'Pert MPD Pro, PANalytical, The Netherlands) to check whether the glasses are amorphous. Thermal properties of the glasses, such as the glass transition temperature (T_g), softening temperature (T_s), crystallization temperature (T_c) and liquid temperature (T_{liq}), were measured using differential thermal analyses for both glasses and glass–ceramics. TECs of glasses and glass–ceramics were determined in stagnant air from room temperature to T_g via dilatometry (L75/1550, LINSEIS, Germany) at a heating rate of 10°C min^{-1}. A total of six specimens were measured for the TEC and the average value was used as the TEC value for a glass. The TEC of 8YSZ was also characterized for the purpose of comparison. The glass–ceramics were characterized by XRD to reveal the precipitated crystalline phases. The wetting behavior of the parent glass on 8YSZ (8 mol% yttria stabilized zirconia) was examined by observing the shape change of a glass pellet with the dimension of 5 × 5 × 1 mm^3 under a home-assembled high temperature microscope, where the glass pellet was put on the top of a 8YSZ plate and continuously heated to 900°C at a heating rate of 5°C min^{-1}. The shape change was recorded using a digital camera connected to the microscope, where pictures were taken at an interval of 10°C.

Long-Term Thermal Stability and Chemical Compatibility

Thermal stability of the as-prepared glasses was investigated through heat treatment at 700°C for up to 500 h. Glasses were first heated to 920°C for 30 min in a Pt crucible with a heating rate of 5°C min^{-1}. They were then cooled with a cooling rate of 5°C min^{-1} down to 700°C and kept for 100–500 h. Glasses were then examined by the XRD to check whether crystallized phases formed after the heat treatment. The as-treated glasses were cut and ground to bars with nominal dimensions of $15 \times 4 \times 2$ mm^3. The bars were subsequently subjected to the TEC characterization using the dilatometer.

The chemical reactivity between the sealing glass and dense 8YSZ was investigated through heat-treatment experiments at 700°C, where a glass/8YSZ couple, with a glass pellet on the top of an 8YSZ disk, was first heated to 900°C for 30 min to ensure a complete wetting of the glass to the 8YSZ disk. It was then cooled down to 700°C and kept for up to 500 h, after which the glass/8YSZ interface was checked using Field-Emission Scanning Electron Microscopy (FESEM, FEI Quanta 200 FEG, Philips, The Netherlands) and the associated energy dispersive X-ray Spectroscopy (EDS) to reveal the chemical reactivity and any microstructural changes.

Bonding and Sealing Tests

In order to test the sealing capability of the as-developed glass, sealing between two metal plates was performed with using a home-made experimental set-up illustrated in Figure 2-3. The metal was SS410 stainless steel (sealing dimension was 50 mm × 50 mm) with a measured TEC of 12.2×10^{-6} K^{-1}(RT-700°C). The glass was first cut and ground to bars of $25 \times 3 \times 1.2$ mm^3 dimensions. The bars were then adhered to the end plate using epoxy glue, followed by vertically putting the upper plate on the top of the end plate. The assembly was subsequently heated to 770°C with a heating rate of 5 K min^{-1} and kept for 30 min to ensure a good sealing before cooling down to 700°C with a cooling rate of 2 K min^{-1}, after which air was introduced into the inner chamber of the setup to a pressure about 5.5–6.5 kPa above the atmospheric pressure. The heating and cooling profiles are shown in Figure 2-4. The change of the pressure was measured with U-tube pressure gauge using water as the pressure measuring

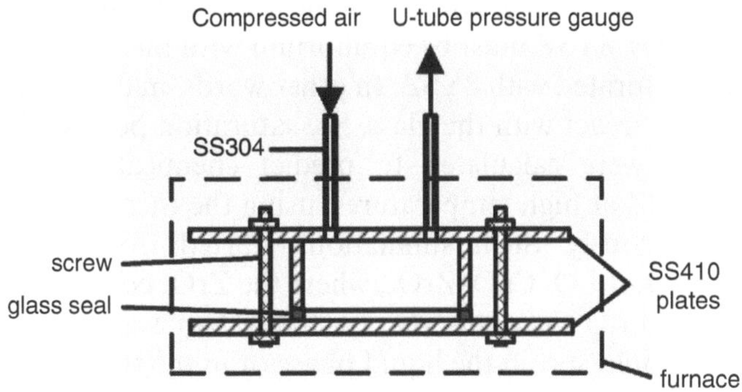

Figure 2-3. Schematic drawing of the leak rate test fixture

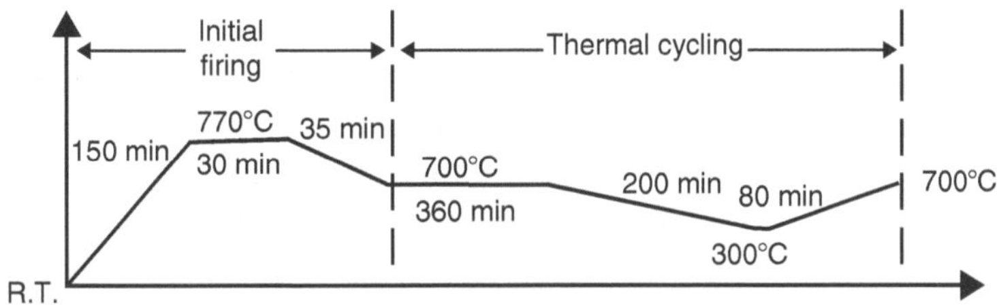

Figure 2-4. Typical temperature profile for initial heat treatment and subsequent thermal cycling tests

media. The leak rate could be calculated based on pressure difference using the ideal gas law as:

$$L = \frac{dP \times 9.80665 \times V}{8.314 \times T} \times 0.0224, \quad (11)$$

where T is the temperature, V is the volume of the inner chamber and equals to 29.7 cm^3 for the current setup, the dP is the instantaneous pressure drop at a inner–outer pressure difference P, L is the leak rate normalized to the standard state in sccm. The calculated leak rate was further normalized with respect to the leak length (20 cm).

RESULTS AND DISCUSSION

Glass Development

During the initial stage of sealing glass development, we put much effort on the chemical compatibility issue, aiming at developing a sealing glass that could be chemically compatible with 8YSZ. According to the thermodynamic theory, if 8YSZ is chemically compatible with a glass at high temperatures, 8YSZ must be equilibrium with the glass, which would need the glass saturated with 8YSZ. In other words, in the saturation state, 8YSZ should not react with the glass. So, saturation points of zirconia in various glasses were calculated to predict chemical compatibility of glasses with 8YSZ at high temperatures, using the thermodynamic model established previously. Such simulations started from the system of SiO_2–B_2O_3–Al_2O_3–Li_2O–CaO–ZrO_2, where the ZrO_2 content in the (56–x) SiO_2–23B_2O_3–15Li_2O–7CaO–$x$$ZrO_2$ (wt%) system was increased to see whether ZrO_2 would stay in the liquid phase or in the solid phase to determine the solution limit of ZrO_2 in the glass system. Simulations in this system showed surprisingly good agreement with those experimental results. For example, the saturation point of zirconia in the glass system was calculated to be around 5 wt%. In comparison, it was determined experimentally to be around 7 wt%. Similar good agreement between the calculated results and experimentally determined values was also demonstrated for the saturation point of alumina in a similar system [32], suggesting that the thermodynamic model was reasonably correct.

Since alkali metal oxides like Li_2O has been proven to be too reactive to various SOFC materials at high temperatures, our sealing glass development focused mainly on glasses without alkali metal oxides. Thermodynamic simulations were subsequently focused on the SiO_2–B_2O_3–BaO–La_2O_3–Y_2O_3–ZrO_2 system. Investigations revealed that simulation results were not as good as those of the SiO_2–B_2O_3–Al_2O_3–Li_2O–CaO–ZrO_2 system, due primarily to the lack of sufficient associate species from ZrO_2, Y_2O_3, La_2O_3, and BaO in the thermodynamic model. We shall add more associate species in our future work to make the model more reliable. However, even with this tentative model, thermodynamic simulation could provide very useful information over glass–8YSZ interactions at high temperatures. For example, the reaction behavior of 8YSZ with glasses was largely dependent on glass composition. Thermodynamic simulations predicted that $ZrSiO_4$ and $BaZrO_3$ would be the main resultant phases after 1 mol glass reacts with 1 mol 8YSZ, which was in a good agreement with our experimental results, where the glass in powder form was mixed with 8YSZ powder and heat treated at 800°C for 100 h. Figure 2-5 shows the XRD pattern of the mixture after the heat treatment, which shows clearly the formation of the Ba_2ZrO_4 phase.

We also intended to investigate thermal stability of a glass through thermodynamic simulation. The purpose was to investigate whether it was possible to avoid the formation of some low TEC phases like $ZrSiO_4$ (4.2×10^{-6} K^{-1},100–200°C) [33], celsian phase (2.29×10^{-6} K^{-1} for 20–300°C) and hexacelsian (7.1×10^{-6} K^{-1} for 20–1,000°C) [12], through manipulating the

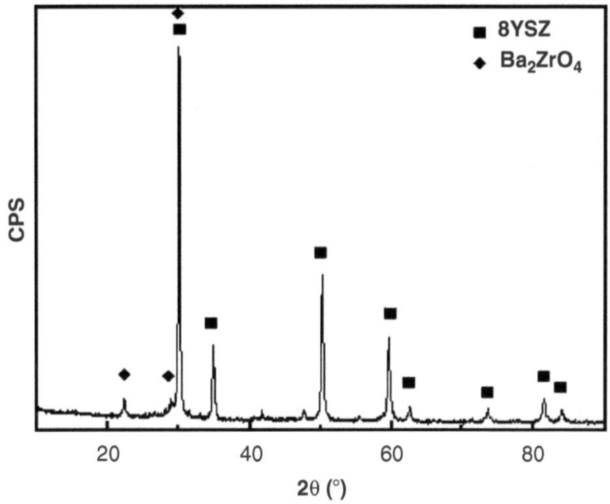

Figure 2-5. XRD pattern of the glass–8YSZ mixture heat treated at 800°C for 100 h in air

glass composition. However, such idea has not been fully realized due to the lack of thermodynamic data for phases like $ZrSiO_4$ and $BaAl_2Si_2O_8$. Nevertheless, our investigations did show that through manipulating the glass composition these low TEC phases could indeed be eliminated, which could significantly improve the thermal stability of the glass. For example, in the glasses containing a high quantity of BaO, $BaAl_2Si_2O_8$ was found to be easily precipitated out from glasses and the successive precipitation of $BaAl_2Si_2O_8$ from these glasses would cause a continuous decrease of TEC value of the glass, which greatly reduces the thermal stability of the glass, as already demonstrated by Sohn et al. [9]. So, the prevention of $BaAl_2 Si_2O_8$ formation is crucial for improving thermal stability of sealing glasses containing a high amount of BaO. Our analyses revealed that the formation of $BaAl_2Si_2O_8$ phases could be avoided if the amount of alumina was limited in the SiO_2–B_2O_3–Al_2O_3– CaO–BaO–Y_2O_3–ZrO_2 system. Subsequent experiments confirmed this point, where glasses with different Al_2O_3 contents were heat treated at 800°C for 10–300 h, followed by XRD characterizations to reveal crystalline phase development. As illustrated in Figure 2-6, for glasses containing 10 wt% Al_2O_3, $BaAl_2Si_2O_8$ phases were easily identified after being heat treated for just 10 h at 800°C. The thermal stability of glasses increases with decreasing Al_2O_3 contents in the glass, where no peaks from $BaAl_2Si_2O_8$ phases could be detected by the XRD characterization after being heat treated for 300 h at 800°C when the glass contained only 1 wt% alumina.

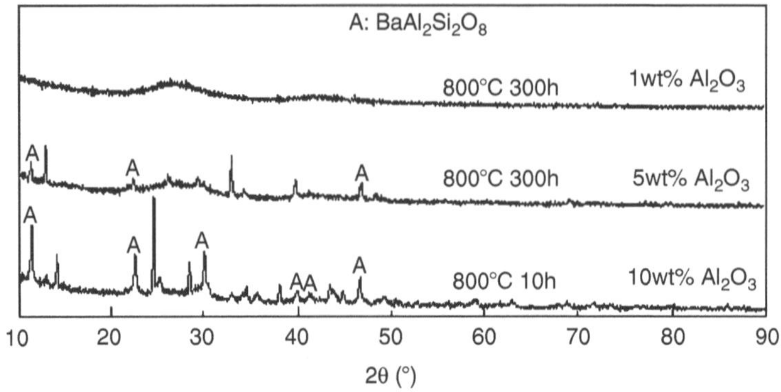

Figure 2-6. XRD patterns of glasses after being heat treated at 800°C for 10–300 h

It should be noted that at the present stage it would be quite difficult to predict complete reaction behaviors for real sealing glass system, using the thermodynamic model previously established. However, the above-mentioned investigations have clearly demonstrated that thermodynamic analyses could provide not only the useful information about various reactions among different materials, but also help to understand glass crystallization behaviors, which could help to find ways to improve sealing glass performances. Further addition of more associate species should make the model more robust for the prediction of various reactions between glasses and other SOFC materials.

After extensive searching in the SiO_2–B_2O_3–Al_2O_3–CaO–BaO–Y_2O_3–ZrO_2 system, a glass with promising properties was successfully developed. The composition of the glass is listed in Table 2-3.

Thermal Properties

Figure 2-7 shows a typical thermal expansion curve for the newly developed glass together with that of 8YSZ, where the TEC value of the glass was calculated to be 9.9×10^{-6} K^{-1} between room temperature (RT) and 631°C (dilatometer determined T_g), very close to the TEC value of 10.0×10^{-6} K^{-1} measured for 8YSZ in the same temperature range. The glass transition temperature is lower than the targeted use temperature (e.g., 700°C), which would be beneficial to thermal stress release at the

Table 2-3. Composition of the newly developed sealing glass

Substance	**SiO_2**	**B_2O_3**	**BaO**	**Additives (ZrO_2, Y_2O_3, La_2O_3, ZnO)**
wt%	24	20	50	6.0

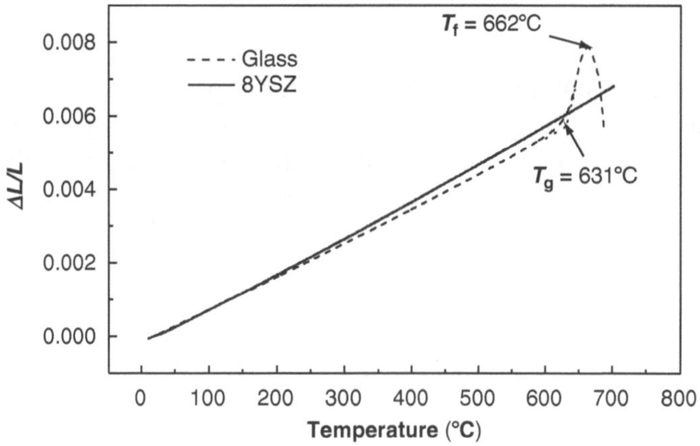

Figure 2-7. Thermal expansion behaviors of a newly developed sealing glass and 8YSZ (10°C min^{-1}, in air)

operating temperature of SOFC since significant stress begin to develop only as the temperature drops below T_g [8].

In order to find the sealing temperature, the wetting behaviors of the newly developed glass on 8YSZ were investigated by observing shape changes of the glass sample on the 8YSZ plate with increased temperature, as illustrated in Figure 2-8. From which, three temperatures 770°C, 810°C and 840°C, corresponding to three characteristic shapes, were determined. At 770°C, the edges of the sample become round. At 810°C, the sample was completely round except the part in contact with the 8YSZ (the ball point temperature), while at 840°C, the shape changed to a hemisphere (called the hemisphere point temperature [34]), and a good wetting between the glass and 8YSZ was achieved with the contact angle less than 90°. Two features could be determined by these observations. Firstly, the glass shows sufficient rigidity at 700°C, which makes it possible to be used as the sealing materials for SOFCs operated at 700°C. Secondly, the sealing temperature can be determined between 810 and 840°C.

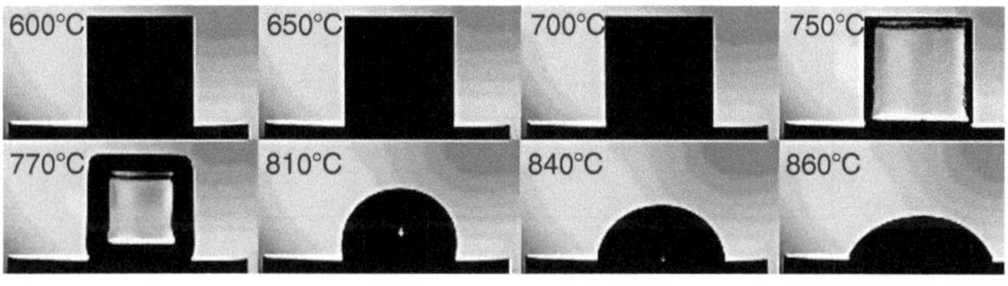

Figure 2-8. Shape change of the glass specimen with increasing temperatures (heating rate 5°C min^{-1})

Thermal Stability

The thermal stability of the newly developed glass was investigated through annealing the glass at 700°C and at 800°C for up to 500 h, after which the TECs were measured and compared with those of the original glass and those reported in literature [9]. Figure 2-9 shows the TEC values against the annealing time. It is seen that after annealing at 800°C for 300 h and at 700°C for 500 h, the TEC change of the glass is less than 2%, which is well within the measuring limit (±1.5%) of the equipment, demonstrating superior thermal stability of the glass. Also, after the annealing tests, no crystalline phases were identified by the XRD characterization and the glass looks completely transparent as demonstrated in Figure 2-10.

Achieving long-term thermal stability is one of the main issues in developing a suitable sealing glass and glass–ceramic for advanced planar SOFC stacks. However, glasses are susceptible to devitrification at high temperatures, which causes significant change of thermal properties. As illustrated in Figure 2-9, the TEC change can be as high as 35% after a similar heat treatment at 800°C for 500 h for the $BaO–Al_2O_3–B_2O_3–SiO_2–La_2O_3–ZrO_2$ glasses reported by Sohn et al. [9]. The present investigation demonstrates, however, that it is possible to develop glasses with superior thermal stability at high temperatures, although further tests are necessary to reveal the long-term thermal stability of the newly developed sealing glass.

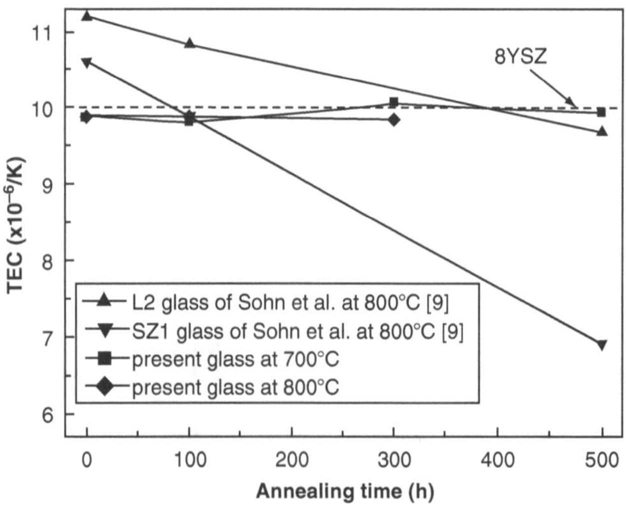

Figure 2-9. Variation of TEC values as a function of annealing time at 700 and 800°C

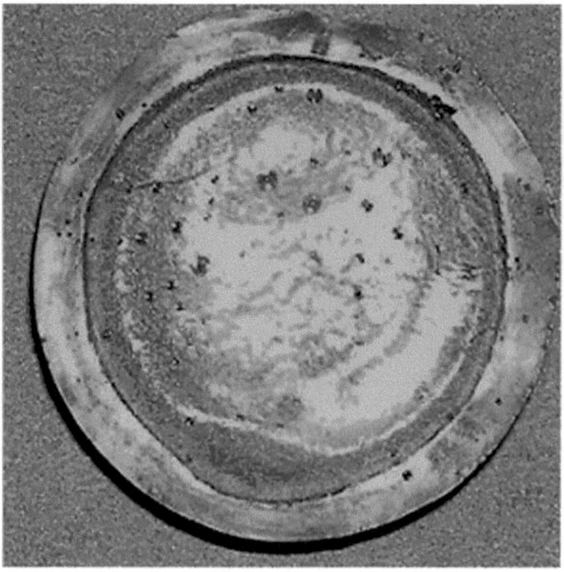

Figure 2-10. Appearance of the glass after being annealed at 700°C for 100 h

Chemical Compatibility with 8YSZ

For SOFC applications, it is quite important to avoid adverse interfacial reactions between sealing materials and the materials to be sealed. The chemical compatibility of the newly developed glass with 8YSZ was therefore examined through heat-treatment experiments at 700°C. Figure 2-11 shows microstructures of the glass/8YSZ interface after the heat-treatment tests, from which it is clear that the glass bonds 8YSZ very well without any micro- and macrocracks. The microstructure shows a clear 8YSZ/glass interface, indicating that interfacial reactions are insignificant after being heat treated with 8YSZ at 700°C for up to 500 h. To further reveal the interfacial interaction, elemental distribution profile was acquired by the EDS line scanning technique and is shown in Figure 2-12. It shows that the concentrations of Zr and Y drop sharply at the interface and the diffusion layer was roughly determined to be 7.6 μm and 8.1 μm after annealing for 100 h and 500 h, respectively, indicating that the interfacial diffusion is limited to a narrow region.

Crystallization Control

Since glasses are thermodynamically unstable, they will eventually crystallize to form glass–ceramics at high temperatures. It is therefore advantageous to control the crystallization process in order to adjust glass

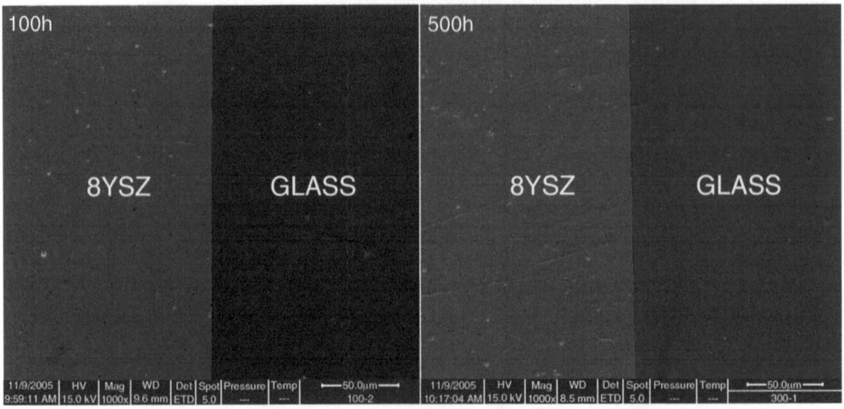

Figure 2-11. Microstructure of the glass/YSZ interface after being heat treated at 700°C for 100 and 500 h

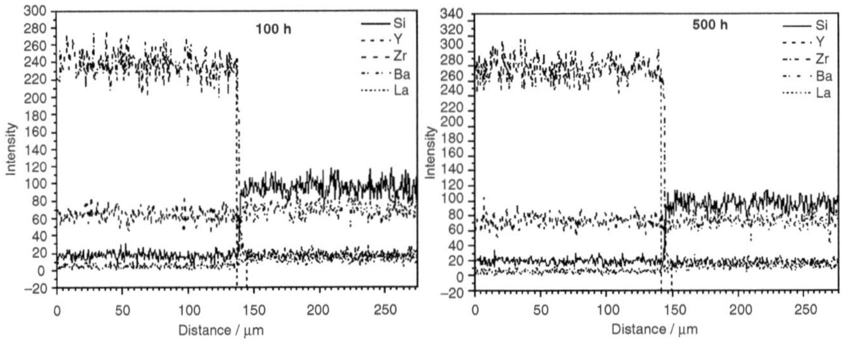

Figure 2-12. EDS line analyses for the glass/YSZ interface heat treated at 700°C for 100 or 500 h

properties. The purposes of the crystallization control include avoiding unwanted phases formation, promoting desirable phases formation, facilitating uniform crystal formation, etc. The crystallization process is usually controlled through manipulating crystallization kinetics (nucleation agents, nucleating conditions, crystallization conditions, etc.).

Our investigations revealed that without a nucleating agent, the newly developed glass is difficult to crystallize. Consequently, various oxides including ZrO_2, TiO_2, Cr_2O_3, P_2O_5 were tried as the nucleating agents for the glass. It was found that the addition of P_2O_5 was quite effective for the crystallization control. The crystallization conditions (e.g., nucleating and crystallization temperatures) were normally determined according to thermal analyses, where it was generally proposed that the nucleation should be performed at a temperature between T_g and 30°C above T_g, while the crystallization temperature should be around the crystallization temperature (T_c) and 30°C below the liquid temperature (T_{liq}). The DSC curve was thus characterized for the newly developed glass with an addition of 5wt% of P_2O_5, as illustrated in Figure 2-13.

The T_g, T_c, and T_{liq} were determined from Figure 2-13 to be 663°C, 791°, 1061°C, respectively. Accordingly, the nucleating temperature was determined to be 692°C and the crystallization temperature was determined to 820–920°C. Figure 2-14 shows the XRD patterns for glasses after the heat treatment under different conditions. It shows that $Ba_{4.5}La_{4.5}O_{24}P_{1.5}Si_{4.5}$ precipitates out as the main crystalline phase after being heat treated at 820°C for 10 h, while $Ba_3LaO_{12}P_3$ is the main crystalline phase for the heat treatment at 880 and 920°C for 10 h, demonstrating that crystallization behavior could be manipulated through adjusting the crystallization kinetics. It is envisaged that the precipitation of $Ba_3LaO_{12}P_3$ would be better since the glass former (SiO_2) is not precipitated out, so the remaining glass would have a higher SiO_2 content and consequently better stability as compared with the parent glass. The TECs were also characterized for the as-crystallized glass. As illustrated in Figure 2-15, an increase in the TEC was recorded for the glass crystallized at 920°C for 10 h, and the as-crystallized glass and 8YSZ have nearly the same thermal expansion behavior before the glass transition temperature.

Although further detailed investigations are definitely needed to obtain a glass ceramic sealing material, the above preliminary investigation demonstrated that the properties of the newly developed glass can be well adjusted through the crystallization control process.

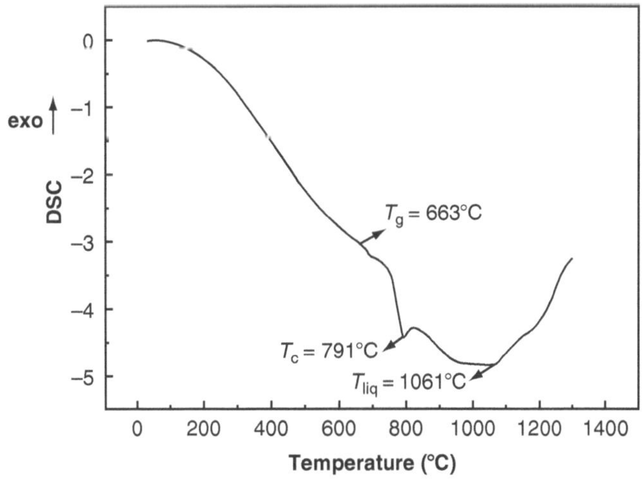

Figure 2-13. DSC curve of the glass with P_2O_5 nucleating agent at a heating rate of 10°C min^{-1} in air

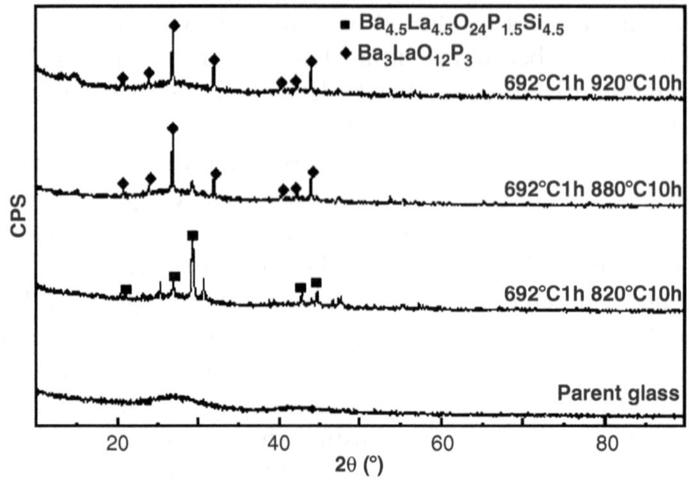

Figure 2-14. XRD of the glass after different heat treatment together with P_2O_5 as the nucleating agent

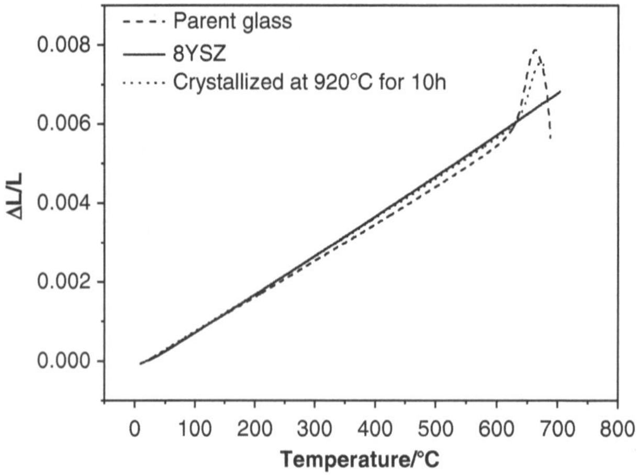

Figure 2-15. Thermal expansion behaviors of parent glass, as-crystallized glass–ceramic and 8YSZ

Sealing Capability

Bonding tests revealed that the sealing temperature was strongly dependent on the load during the sealing process, e.g., without a load, a temperature over 830°C would be necessary to achieve a good wetting and bonding between the glass and 8YSZ or metal interconnect. However, when applying a load of 8.56 kPa (weight of the sealing test fixture), the sealing temperature could be reduced to 770°C. On the other hand, good sealing cannot be achieved below 750°C, irrespective the load applied. Investigations revealed that bonding strength between the glass and 8YSZ or SS410 stainless steel was quite good, e.g., after intentional breaking the sealing, cracks were actually found to propagate within the glass matrix, not along the glass–metal interface, indicating that the interfacial bonding strength is higher than the strength of the glass.

The sealing capability of the newly developed glass was subsequently tested through measuring the leak rate of air sealed in the chamber shown in Figure 2-3. Before real tests at 700°C, background leak rate was first tested at room temperature using a rubber seal. The background pressure drop with time is illustrated in Figure 2-16, where the figure also reveals that the leak rate is dependent on the pressure difference and it was calculated to be 1.38×10^{-5} sccm cm^{-1} at a pressure difference of 1.4 kPa. Ideally, the leak rate should be zero for a hermetic seal. In reality, the actual low leak rates were limited by the system's background since there were valves and tube connectors in the setup [35].

Figure 2-17 shows the leak rates vs. thermal cycle numbers, where the leak rate of the as-sealed setup is 2.63×10^{-5} sccm cm^{-1}, which is twice as measured background leak rate. Interestingly, the leak rate decreases with increasing the number of thermal cycles. As illustrated in Figure 2-17, after three thermal cycles, the leak rate decreased to 9.76×10^{-6} sccm cm^{-1}, which is even lower than the background leak rate. The reason for this might be due the fact that the background leak rate was measured using a rubber seal at room temperature, which may be different significantly as compared with that of glass seal measured at 700°C.

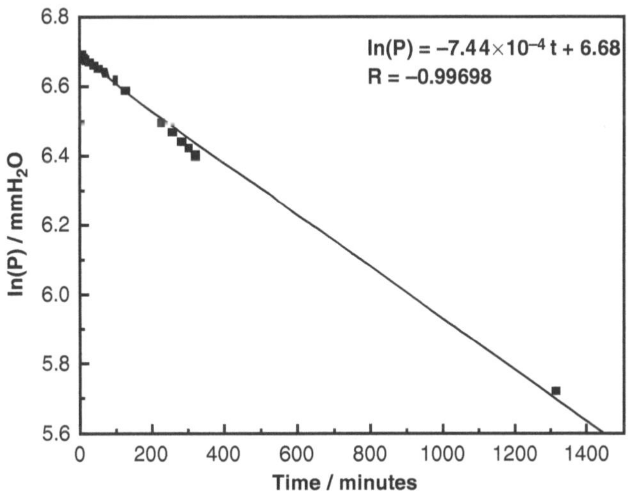

Figure 2-16. Pressure drop of the background leak test for the current test setup

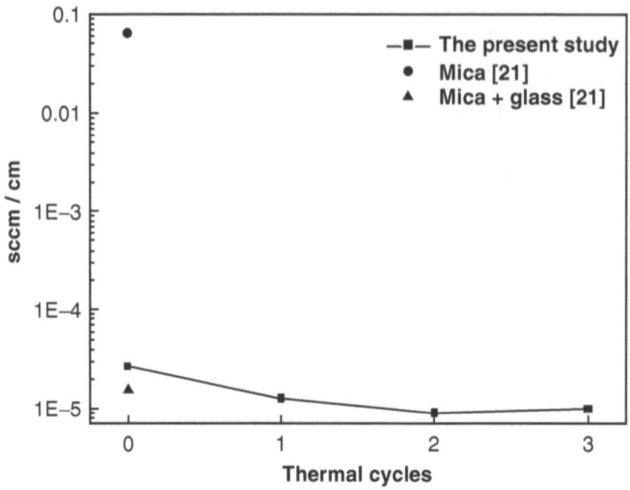

Figure 2-17. Normalized leak rates (sccm cm^{-1}) vs. the number of thermal cycles

The leak rates reported by Chou et al. [21] on mica-based seals were normalized to the pressure of 1.4 kPa and are plotted in Figure 2-17 for comparison. It is clear that the leak rate is nearly the same as reported for the mica–glass composite compressive seals in the literature [21], and it is three orders of magnitudes lower than that of the pure mica compressive seal [21]. The above preliminary sealing tests demonstrated that very good hermetic capability could be achieved for the newly developed sealing glass, together with the capability of withstanding thermal cycles.

CONCLUSIONS

In the present study, a new approach, aiming at quasiquantitative design of sealing glasses, has been proposed. By this method, a glass was developed for SOFCs operated at 700°C. The glass has a thermal expansion coefficient of 9.9×10^{-6} K^{-1}, which is quite close to the value of 10.0×10^{-6} K^{-1} measured for 8YSZ under similar conditions. The sealing glass exhibits superior long-term thermal stability, where after being heat treated for 500 h at 700°C, the change of the TEC value is within the range of equipment error. The glass shows good chemical compatibility with 8YSZ, where no obvious interfacial reaction was detected after being heat treated for 500 h at 700°C together with 8YSZ. Also, the glass properties can be adjusted through the control of crystallization process to obtain a suitable glass–ceramic.

The newly developed glass can wet and bond 8YSZ and SS 410 stainless steel readily above 830°C and it can be effective sealed with 8YSZ and SS410 stainless above 770°C with the applying load of 8.56 kPa. Sealing tests demonstrated that the glass has very good hermetic capability, where the leak rate was measured to be 9.76×10^{-6} sccm cm^{-1}.

The newly developed sealing glass could be a promising candidate for advanced planar SOFC applications.

ACKNOWLEDGEMENTS

The authors would like to acknowledge the financial support from the National Natural Science Foundation of China(Grant No. 20406023).

References

[1] O. Yamamoto, Solid oxide fuel cells: fundamental aspects and prospects, *Electrochimica Acta*, 45 (2000) 2423.

[2] P. Costamagna, A. Selimovic, M.D. Borghi, and G. Agnew, Electrochemical model of the integrated planar solid oxide fuel cell (IP-SOFC), *Chemical Engineering Journal*, 102 (2004) 61.

[3] P. Singh, Nasa, Pacific northwest team on SOFC sealing, *Fuel Cells Bulletin*, 2 (2004) 6.

[4] W. Fergus, Sealants for solid oxide fuel cells, *Journal of Power Sources*, 147 (2005) 46–57.

[5] K.D. Meinhardt, J.D. Vienna, T.R. Armstrong, et al., Glass–ceramic material and method of making, US Patent No. 6430966, August 13, 2002.

[6] P.H. Larsen, C. Bagger, M. Mogensen, et al., *Proceedings of Fourth International Symposium on Solid Oxide Fuel Cells*, eds. M. Dokiya, O. Yamamoto, H. Tagawa and S.C. Singhal The Electrochemcial Society, Pennington, NJ, V95-1, 1995, pp. 69–78.

[7] P.H. Larsen, F.W. Poulsen, and R.W. Berg, The influence of SiO_2 addition to $2MgO–Al_2O_3–3.3P_2O_5$ glass, *Journal of Non-Crystalline Solids*, 244 (1999) 16–24.

[8] K. Ley, M. Krumpelt, R. Kumar, et al., Glass–ceramic sealants for solid oxide fuel cells: Part I. Physical properties, *Journal of Materials Research*, 11 (1996) 1489–1493.

[9] S.B. Sohn, S.Y. Choi, G.H. Kim, et al., Suitable glass–ceramic sealant for planar solid-oxide fuel cells, *Journal of the American Ceramic Society*, 87 (2004) 254–260.

[10] P.H. Larsen and P.F. James, Chemical stability of $MgO/CaO/Cr_2O_3–Al_2O_3–B_2O_3$–phosphate glasses in solid oxide fuel cell environment, *Journal of Materials Science*, 33 (1998) 2499–2507.

[11] Z. Yang, J.W. Stevenson, and K.D. Meinhardt, Chemical interactions of barium–calcium–aluminosilicate-based sealing glasses with oxidation resistant alloys, *Solid State Ionics*, 160 (2003) 213–225.

[12] K. Eichler, G. Solow, P. Otschik, et al., BAS ($BaO–Al_2O_3–SiO_2$)-glasses for high temperature applications, *Journal of the European Ceramic Society*, 19 (1999) 1101–1104.

[13] C. Lara, M.J. Pascual, and A. Duran, Glass-forming ability, sinterability and thermal properties in the systems $RO–BaO–SiO_2$ (R=Mg, Zn), *Journal of Non-Crystalline Solids*, 348 (2004) 149–155.

[14] I.D. Bloom and K.L. Ley. Compliant sealants for solid oxide fuel cells and other ceramics, US Patent No. 5453331, Sept. 26, 1995.

[15] V.A.C. Haanappel, V. Shemet, S.M. Gross, et al., Behaviour of various glass–ceramic sealants with ferritic steels under simulated SOFC stack conditions, *Journal of Power Sources*, 150 (2005) 86–100.
[16] M.C. Tucker, C.P. Jacobson, L.C. De Jonghe, et al., A braze system for sealing metal-supported solid oxide fuel cells, *Journal of Power Sources*, 160(2) (2006) 1049–1057.
[17] K.S. Weil, C.A. Coyle, J.T. Darsell, et al., Effects of thermal cycling and thermal aging on the hermeticity and strength of silver–copper oxide air-brazed seals, *Journal of Power Sources*, 152 (2005) 97–104.
[18] M. Bram, S. Reckers, and P. Drinovac, Characterization and evaluation of compression loaded sealing concepts for SOFC stacks, *Electrochemical Society Proceedings*, 7 (2003) 888–897.
[19] S.P. Simner and J.W. Stevenson, Compressive mica seals for SOFC applications, *Journal of Power Sources*, 102 (2001) 310–316.
[20] Y.S. Chou and J.W. Stevenson, Thermal cycling and degradation mechanisms of compressive mica-based seals for solid oxide fuel cells, *Journal of Power Sources*, 112 (2002) 376–383.
[21] Y.S. Chou, J.W. Stevenson, and L.A. Chick, Novel compressive mica seals with metallic interlayers for solid oxide fuel cell applications, *Journal of the American Ceramic Society*, 86(6) (2003) 1003–1007.
[22] M. Cable, Classical glass technology. *Materials Science and Technology, Vol. 9: Glasses and amorphous materials*, eds. R.W. Cahn, P. Haasen, and E.J. Kramer. VCH, Weinheim, Germany, 1991, pp. 30–31.
[23] T. Lakatos, L.-G. Johansson, and B. Simmingskold, Viscosity temperature relations in the glass systems, *Glass Technology*, 13 (1972) 88–95.
[24] H. Zhang and Q. Zhu, Glass composition prediction using a combined GA-BFGS algorithm, *Computer and Applied Chemistry*, 20 (2003)335–339.
[25] W.B. White, S.M. Johnson, and G.B. Dantzig, Chemical equilibrium in complex mixtures, *The Journal of Chemical Physics*, 28 (1958) 751.
[26] T.M. Besmann and K.E. Spear, Thermochemical modeling of oxide glasses, *Journal of the American Ceramic Society*, 85 (2002) 2887–2894.
[27] A.D. Pelton and M. Blander, Thermodynamic analysis of ordered liquid solutions by a modified quasi-chemical approach-application to silicate slags, *Metallurgical Transactions B*, 17B (1986) 805–815.
[28] A.D. Pelton and P. Wu, Thermodynamic modeling in glass-forming melts, *Journal of Non-Crystalline Solids*, 253 (1999) 178–191.
[29] K.E. Spear, T.M. Besmann, and E.C. Beahm, Thermochemical modeling of glass: application to high-level nuclear waste glass, *MRS Bulletin* 24 (1999) 37–44.
[30] I. Barin, F. Sauert, E. Schultze-rhonhof, et al., *Thermochemical Data of Pure Substance*, Federal Republic of Germany, 1993.
[31] M.W. Chase Jr., C.W. Davies, J.R. Downey Jr., et al., JANAF Thermochemical Tables, 3rd ed. *J. Phys. Chem. Ref. Data*, Vol. 14, Suppl. 1. American Chemical Society, New York, 1985.
[32] Q. Zhu, G. de With, L.J.M.G. Dortmans, et al., Near net-shape fabrication of alumina glass composites, *Journal of the European Ceramic Society*, 25 (2005) 633–638.
[33] I.W. Donald, Preparation, properties and chemistry of glass and glass–ceramic-to-metal seals and coatings, *Journal of Materials Science*, 28 (1993) 2841–2886.
[34] T. Schewickert, R. Sievering, and P. Geasee, Glass–ceramic materials as sealants for SOFC applications, *Materialwiss Werkst*, 33(6) (2002) 363–366.
[35] Y.S. Chou, J.W. Stevenson, and L.A. Chick, Ultra-low leak rate of hybrid compressive mica seals for solid oxide fuel cells, *Journal of Power Sources*, 112 (2002) 130–136.

Chapter 3

A NOVEL TECHNOLOGY OF SOLID OXIDE FUEL CELL FABRICATION

Piotr Jasinski[1], Toshio Suzuki[2], Vladimir Petrovsky[3], and Harlan U. Anderson[3]
[1]Gdansk University of Technology, Faculty of Electronics, Telecommunication and Informatics, ul. Narutowicza 11/12, Gdansk 80-952, Poland, [2]National Institute of Advanced Industrial Science and Technology, AMRI, 2266-98 Anagahora Shimo-Shidami, 463-8560 Nagoya, Japan, [3]University of Missouri Rolla, Electronic Materials Applied Research Center, 303 MRC, 65409 Rolla, USA

Solid oxide fuel cells (SOFCs) are one of the most efficient energy conversion devices [1]. The main demand in the current SOFC development is lowering operation temperature to the range of 600–800°C – intermediate temperature SOFC (IT-SOFC). In order to lower operational temperature and increase or at least sustain performance comparable to that at high temperature SOFCs, it is necessary to decrease the resistance of the electrolyte and lower the overpotential of the electrodes. One of the ways to achieve this goal is to decrease the thickness of the electrolyte and optimize the structure of the electrodes.

Several techniques can be used for the fabrication of SOFCs. The methods include chemical and physical deposition, such as electrochemical vapor deposition (EVD) [2] or magnetron sputtering [3], and liquid precursor and powder processing techniques such as polymer spin coating [4], die pressing, screen printing, or tape casting [5]. All of these techniques experience difficulties with deposition of 1–10 μm thick dense electrolyte layers. Moreover, the methods used for electrode fabrication frequently cannot comply with the requirements related to porosity of layer and composition of mixed conductor composite structure. The net shape technology is the method which seems to overcome those limitations.

This chapter describes the development of a low temperature method for the deposition of 1–10 μm thick films and bulk materials using a combination of powder processing technologies (colloidal suspension, die pressing, screen printing) and polymeric precursor processing methods (spin coating, impregnation). A schematic diagram for film preparation by a process called net shape processing is presented in Figure 3-1 [6]. In this process the

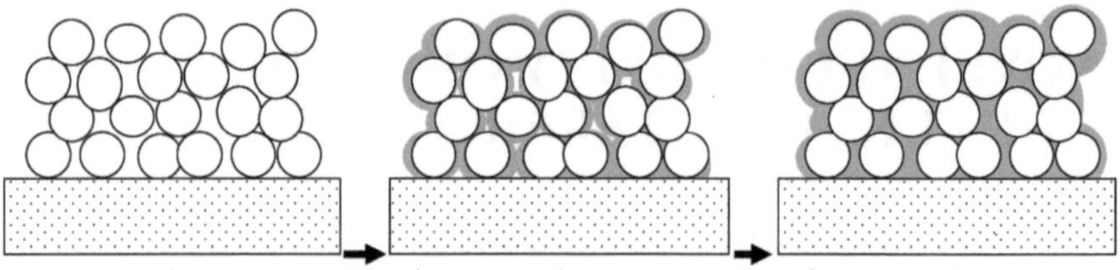

Figure 3-1. The schematic diagram of colloidal-polymer process

powder is first deposited onto the surface to provide a framework of connected particles into which the polymeric precursor impregnates to form an oxide particle/organic polymer composite film. Upon heating to about 300°C, the polymer decomposes to yield a nanocrystalline layer. Subsequent and multiple depositions of polymer precursor fill the space between the particles providing on demand: dense or porous and homogenous or composite structures. The homogenous structures are created when the oxide in the polymer precursor layer is of the same composition as the powder used to create the framework, while heterogeneous composites occur when the two differ. The advantage of this technology is related to low temperature processing, because the interfacial reactions between different layers can be avoided and the problems due to thermal expansion mismatch of different materials can be minimized. The low temperature processing also allows preparation of nanocrystalline structures, which is one of the ways to improve the catalytic properties of electrodes. In addition, the possibility of fabricating dense and nanocrystalline layers at low temperature gives a unique opportunity to investigate nanocrystalline properties of materials, which may be different from those of microcrystalline. Examples of the use of net shape technology for the preparation of the materials typically used in the fabrication of the SOFCs will be shown below.

DENSE ELECTROLYTE FILMS

Lowering the resistance of the electrolyte is one of the objectives important for the development of the performance of IT-SOFCs. To meet this objective the cost effective technology should be developed, which allows preparation of dense electrolyte layer of 1–10 μm thick.

Currently available technologies do not meet those criteria. Powder-based techniques (such as tape casting, screen printing, etc.) are cost effective, but they have limitations such as a thin layer deposition or a layer density. In addition, the required high temperature cosintering of the electrode and electrolyte limits application of these techniques to anode supported SOFCs because of chemical reactions between cathode and electrolyte during high temperature sintering of cathode supported SOFCs. Contrary to powder-based techniques, polymer technologies limit the thickness of one deposition to ~100 nm. During deposition of thicker layers shrinkage connected with decomposition of organic compounds causes cracking of the film. In general, chemical and physical deposition techniques allow deposition of dense electrolyte layers at low temperature, but the cost of such kind of processing is too high for widespread applications [7].

Yttria stabilized zirconia (YSZ) is the most commonly used material for the SOFC electrolyte and several deposition techniques have been used to develop thin film YSZ on either the anode or cathode [7]. To the check applicability of net shape technology for fuel cell fabrication the YSZ films were prepared on Pt and sapphire substrates [8]. A colloidal suspension was prepared using commercially available YSZ powder with an initial grain size of about 100 nm, which was dispersed ultrasonically in water and ethanol. The suspension contained about 50 wt% of YSZ powder. The YSZ polymer was prepared using zirconium chloride and yttrium nitrate precursors. First a few depositions of colloidal suspension and then several depositions of polymer precursor are spin coated on the substrate. The thickness of the film is mainly controlled by number of colloidal suspension depositions, while the ratio between YSZ powder and YSZ derived from polymer precursor controls the final porosity of the material as schematically presented in Figure 3-2. For the present concentration of the suspension and with the spin speed of 1,000 rpm, the thickness of about 1 μm per one deposition of colloidal suspension was obtained. Dense coatings can be prepared when the ratio between YSZ powder and polymer precursor is equal to 1:1. Coatings with controlled porosity can also be prepared using higher ratios of powder to polymer.

Dense (powder and polymer ratio of 1:1) and porous (ratio of 5:1) 1–3 μm thick YSZ films were prepared on sapphire and Pt foil substrates and evaluated. The dense YSZ coatings (powder and polymer ratio of 1:1) appeared to be transparent after deposition (annealing temperature 400°C). The YSZ coatings (powder and polymer ratio of 5:1) were semi-transparent. The coatings on both sapphire and Pt substrates were free of cracks as they were initially made and after the further high temperature annealing (up to 1,000°C). The microstructure of fracture cross sections of

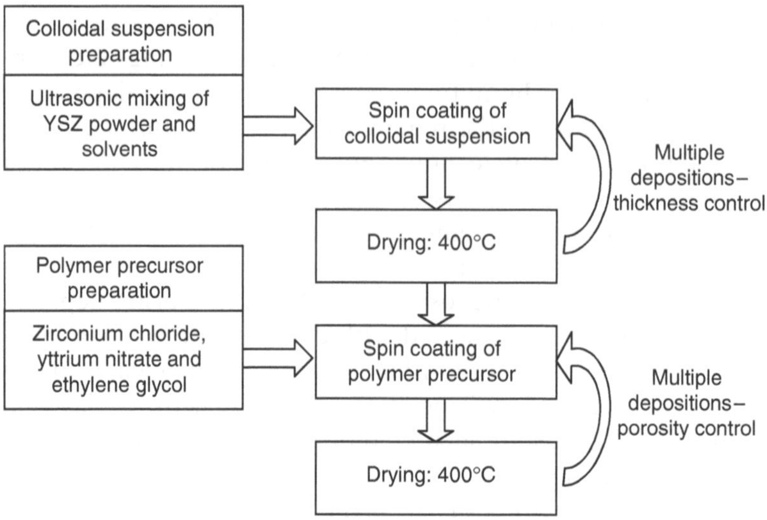

Figure 3-2. Schematic diagram of film preparation by net shape technology

the coatings was examined by scanning electron microscopy (SEM) and summarized in Figures 3-3 and 3-4. Figure 3-3a shows a dense YSZ coating (powder and polymer ratio of 1:1) as prepared on the sapphire (after annealing at 400°C). Two types of grains are observable in this coating: grains of about 100 nm derived from the YSZ powder which formed the framework and 10 nm nanocrystalline YSZ that originated from the polymer decomposition, which fills the space between these grains. When the annealing temperature was increased, the nanocrystalline grains crystallized and created a uniform microstructure. After annealing at 900°C (Figure 3-3b) the crystallization process is completed and only 100–200 nm grains are visible. The important feature is that this recrystallization occurs in a dense structure (more than 95% of theoretical density) and no shrinkage occurred.

Similar processes took place in the case of porous YSZ coating (powder and polymer ratio of 5:1). The only difference is the amount of nanocrystalline YSZ derived from the polymer was not sufficient to fill all space in the initial YSZ framework resulting a porous ceramic material with well-developed connection between particles (Figure 3-3c). This type of material cannot be used as an oxygen separation membrane because it has open porosity, but it can be useful as the YSZ skeleton for the electrodes, as it has high effective surface area for exchange with the gaseous phase.

Figure 3-4 shows a SEM image of a cross section and AFM image of the surface of the dense YSZ film (powder and polymer ratio of 1:1) deposited on platinum foil. It can be seen from the cross-section SEM image (Figure

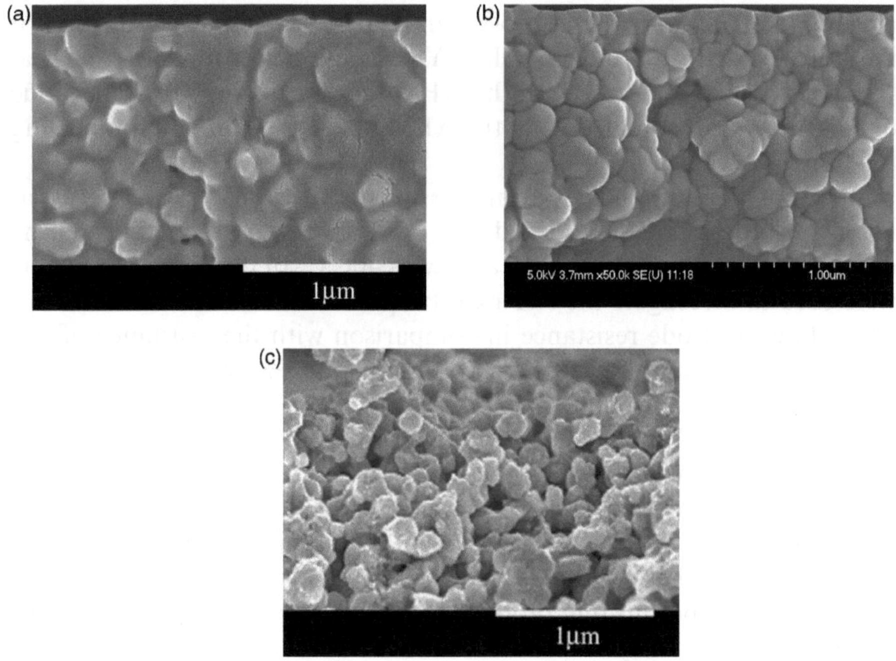

Figure 3-3. SEM images of a fracture cross section of 3 μm thick YSZ film on sapphire: (**a**) powder and polymer ratio of 1:1 sintered at 400°C; (**b**) powder and polymer ratio of 1:1 sintered at 900°C; (**c**) powder and polymer ratio of 5:1 sintered at 900°C

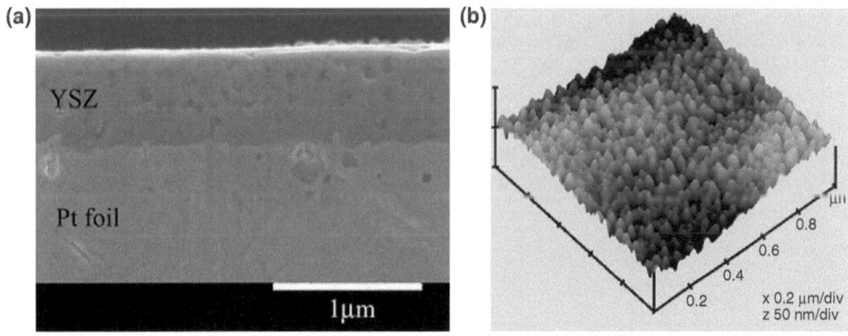

Figure 3-4. SEM image of the fracture cross section (**a**) and AFM image of the surface (**b**) of 1 μm thick YSZ film on Pt foil

3-4a) that the film is well attached to the platinum substrate, with no open porosity and looks similar to the coating on sapphire. An important feature of the net shape processing is the ability to planarize rough surfaces (in this case platinum foil). The coating completely covers the relief of the platinum foil and provides a smooth and glossy surface. The grain size of the polymer-derived nanocrystalline YSZ is the only source of the final roughness which can be seen in the AFM image (Figure 3-4b). The value of this grain size estimated from the AFM image is 50 nm after annealing the film at 800°C.

It is possible to conclude from the microstructural investigation that net shape processing allows the deposition of dense YSZ coatings on the dense substrates at temperatures as low as 400°C. In plane measurements were used to investigate the electrical properties of the structures on sapphire. Low electrode resistance in comparison with the resistance of the material is an advantage in the case of in plane measurements. However, it is impossible to separate the impact of grain and grain boundaries resistances from the material resistance because of the influence of sample holder capacitance. Figure 3-5 shows the temperature dependence of the conductivity of the YSZ films on sapphire. The conductivity of the porous film with powder and polymer ratio of 5:1 is about one order of magnitude lower than that for the dense sample (with the ratio 1:1). The conductivity of this porous film produced at 400°C decreased after heating to 900°C. This is probably related to the densification of the nanocrystalline YSZ derived from the polymer, which causes a decrease in the contact area

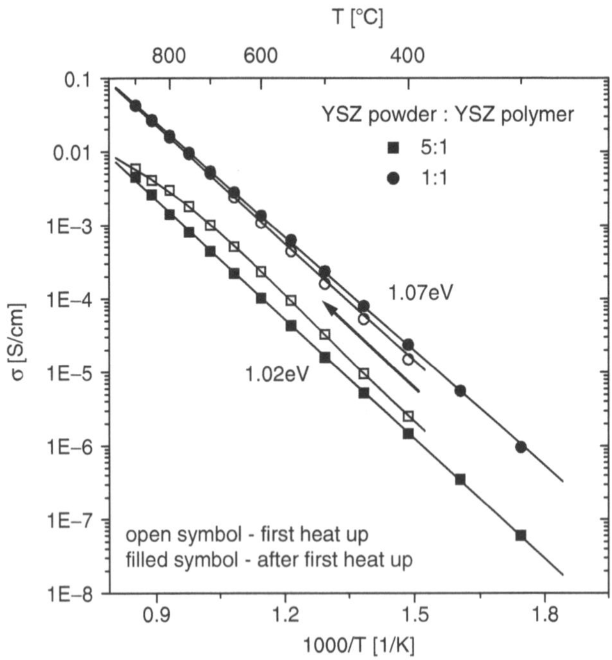

Figure 3-5. Temperature dependence of conductivity of YSZ film on sapphire

between grains in this porous material. In contrast, the conductivity of the dense film produced at 400°C is lower than the conductivity after heating to 900°C, which again can be explained by the crystallization of polymer-derived YSZ. In this case the material is dense and the contact surface area cannot be changed, so crystallization will cause only a decrease of the grain boundary volume. It is known that a decrease of the grain boundary volume results in an increase of the total conductivity of the film [9]. The final conductivity (after short time annealing at 900°C) for both specimens is stable and does not change after the subsequent annealing at 900°C. The conductivity and activation energy for the dense film are very close to those reported for bulk YSZ [10].

The measurements on Pt foil substrate were made using in the thickness configuration, i.e., one of the electrodes was Pt foil, while the other was deposited on the top of the film. In this case more information can be obtained from the impedance spectra, because the grain and grain boundary capacitances are higher than that of the sample holder capacitance, so grain and grain boundary resistances can be separated. The temperature dependences of the grain and grain boundary conductivities were calculated and are presented in Figure 3-6. Below 450°C, grain and grain

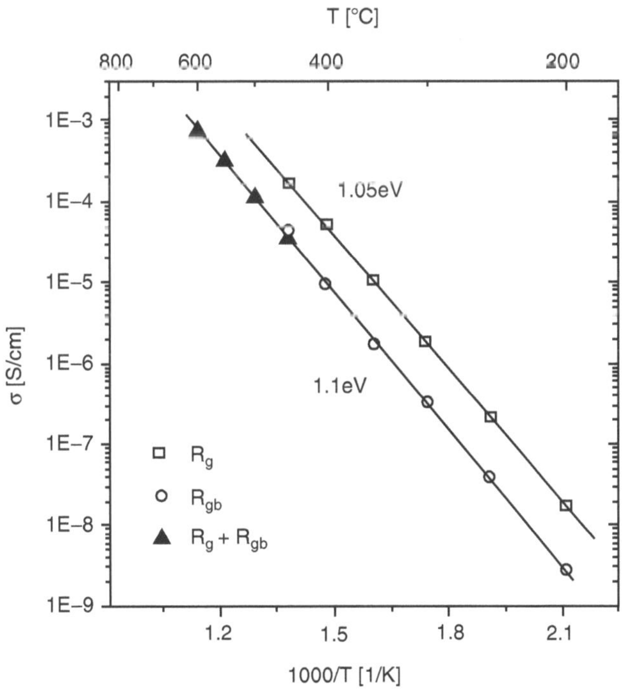

Figure 3-6. Temperature dependence of conductivity of YSZ film on Pt foil

boundary conductivity were plotted individually, while above 450°C the grain and grain boundary conductivity could not be separated so only the total ionic conductivity is presented. Both the activation energies and the values of grain and grain boundary conductivities are in good agreement with the literature data for bulk YSZ [10] and with that measured in plane on the sapphire substrate.

It is interesting to note, that low temperature processing of the net shape prepared YSZ still allows obtaining very high electrical conductivity [6]. Figure 3-7 presents the electrical conductivity dependence of YSZ at 400°C as a function of annealing temperature. The YSZ was prepared by the net shape technology, by the polymer precursor process [11] and by the conventional tape-cast technique. In case of the tape-casted YSZ the same powder was used for the slurry preparation as that for the colloidal suspension. The figure clearly shows that each technique requires different annealing temperature to obtain the same high conductivity. The lowest annealing temperature, as low as 400°C, is required in the case of net shape method.

Doped ceria is another material frequently used in fuel cell fabrication [12]. It is applied as electrolyte and component of electrodes to improve

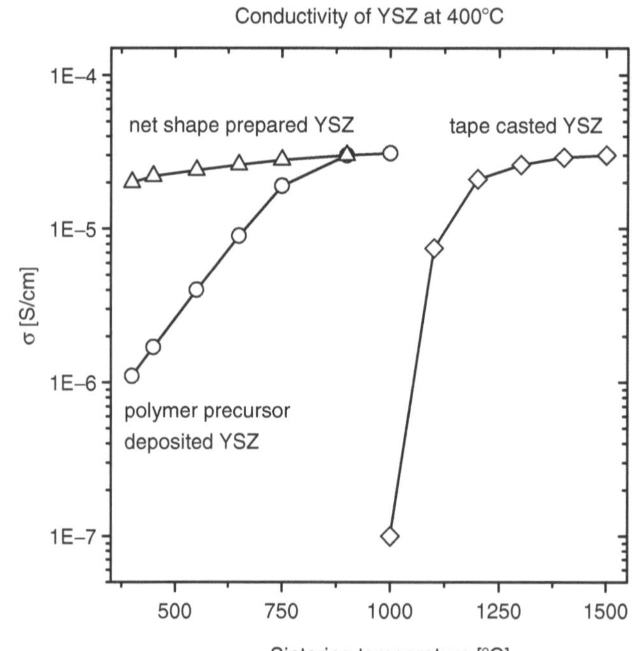

Figure 3-7. The electrical conductivity of the YSZ at 400°C as a function of annealing temperature for several processing techniques

their ionic conductivity and electrocatalytic properties. Gd and Sm are the most commonly used dopants. The net shape process was used for fabrication of 20% Sm-doped ceria (CSO) films [13]. A procedure, similar to the case for YSZ (Figure 3-2), was used to prepare CSO film on Pt foil. A cross section of a ceria film on Pt is shown in Figure 3-8a. The film was 8.2 μm thik and free of any cracks. Figure 3-8b and c are AFM images of the film sintered at 800 and 1,000°C, respectively. The surface of the film shows crystallites of 40 and 55 nm in size for the sintering temperature of 800 and 1,000°C, respectively. An Arrhenius plot of conductivity in the low temperature range is presented in Figure 3-9. As can be seen, the grain boundary conductivity is about one order of magnitude lower than that of the grain. The activation energy was calculated to be of 0.7 and 0.9 eV in case of the grain and grain boundary resistance, respectively. The grain boundary resistance for the 55 nm grain size ceria is about two times higher than that for 40 nm. It can be concluded that the electrical properties of ceria film prepared by net shape technology corresponds to the reported data of nanocrystalline ceria [14].

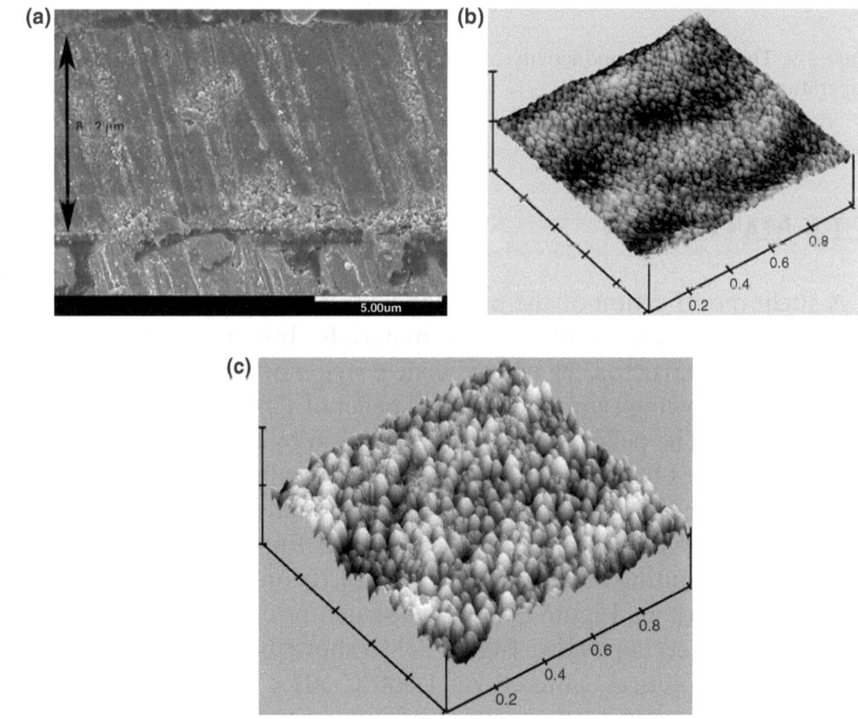

Figure 3-8. The images of Sm-doped ceria film on Pt: the cross section of the film sintered at 800°C (**a**) and AFM images of the film surface sintered at 800°C (**b**) and 1,000°C (**c**)

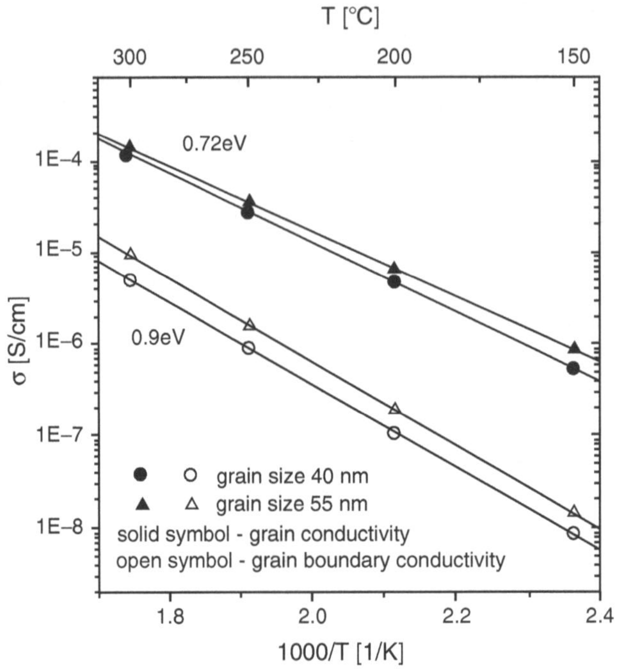

Figure 3-9. The electrical conductivity of Sm-doped ceria film on Pt sintered at different temperatures

BULK MATERIALS

A slight modification of the net shape processing procedure allows the fabrication of nanocrystalline bulk materials. Instead of forming the porous skeleton structure by colloidal suspension other techniques can be used. In the following example a porous pellet of CSO was formed by isostatic pressing. The pellet was then impregnated by CSO polymer precursor multiple times using the procedure schematically presented in Figure 3-10 [13]. After each impregnation the pellets were annealed at 700°C, so that the polymer precursor could decompose. Figure 3-11 shows SEM images of the resulting bulk ceria. The 58% of theoretical density skeleton of ceria is shown in Figure 3-11a. Its structure consisted of fiber-like agglomerates of ceria powder. Figure 3-11b shows bulk ceria with density of 91% of theoretical sintered at 1,000°C after polymeric precursor impregnation. Its structure consisted of nanocrystallites (Figure 3-11b), with grain size about 40 nm. The temperature dependence on the electrical conductivities of the bulk ceria sintered at different temperatures is

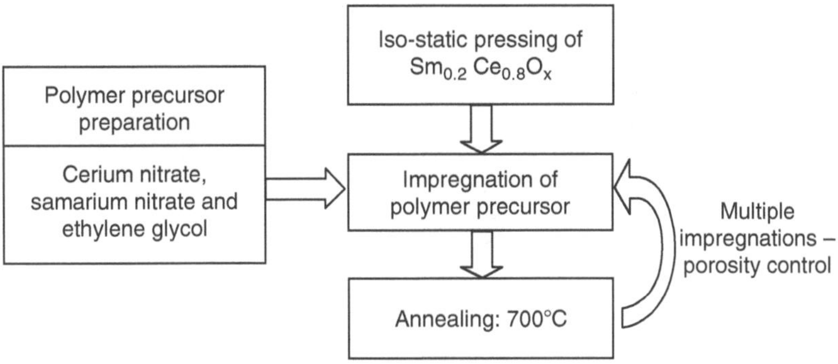

Figure 3-10. Schematic diagram of bulk ceria preparation

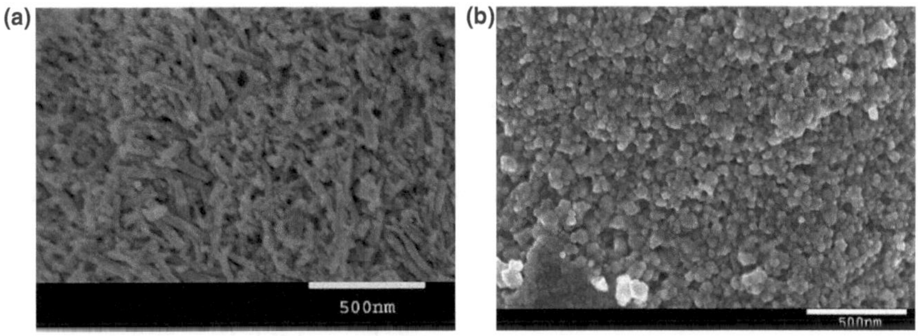

Figure 3-11. SEM image of the fracture of Sm-doped ceria skeleton used for soaking (**a**) and net shape prepared ceramics sintered at 1,000°C (**b**)

presented in Figure 3-12. The ceramics with smaller grains sizes show lower conductivity. This is pronounced in the lower temperature range, what is related to the higher activation energy of the ceria with smaller grains sizes. The electrical conductivity of the ceria, which was sintered at 1,100 and 1,300°C is virtually the same. However, it is lower than that which was measured for die pressed and sintered ceria. Most likely, the lower conductivity of net shaped prepared ceria is related to cracks, which may have resulted from gas evolution from decomposition of the polymer. Therefore, in the case of bulk materials special attention needs to be paid during the decomposition procedure of the polymer to avoid detrimental gas evolution during decomposition of the polymer. Lower conductivity might also be related to the deleterious effects of silicon impurities in the polymer precursor. Although high purity components were used to prepare polymer precursor, the electrical conductivity of ceria can be influenced by small amounts of silicon impurity [15].

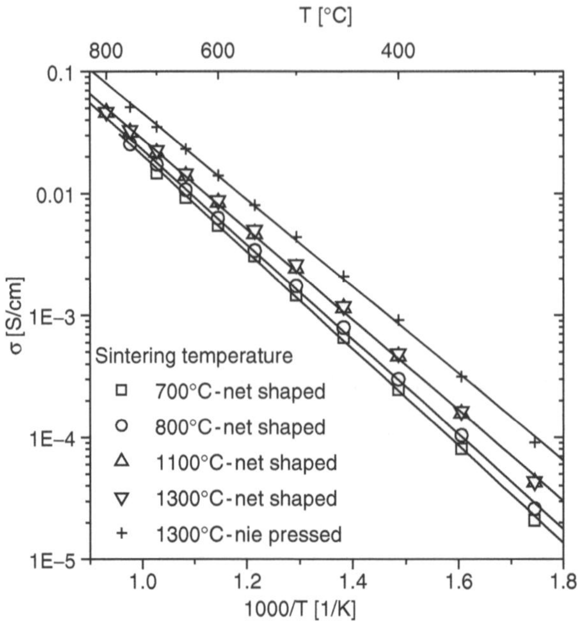

Figure 3-12. The electrical conductivity of Sm-doped ceria ceramics sintered at different temperatures

Another important feature of the net shape technology is that it does not follow percolation theory. This gives an opportunity to design materials with unique properties. As an example Ni–ceria cermet with low Ni content was prepared and investigated [16]. The net shape processed Ni–ceria cermet, which was prepared according to procedure presented in Figure 3-13, features:

- low sintering temperature of backfilled Ni, which allows preparation of nanocrystalline Ni with high surface area and high catalytic activity for reduction of hydrogen,
- high conductivity with as low as 10 vol% of Ni, which reduces the volume change occurring during reduction and oxidation of the anode.

Figure 3-14 shows a SEM image of a Ni-cermet fracture prior to reduction. The image shows a ceria framework with large channels filled with nanocrystalline NiO (Figure 3-14a). The channel structure resulted from the use of the carbon pore former. The grain sizes of the NiO formed in the channels was found to be about 200 nm (Figure 3-14b).

The temperature dependence of the electrical conductivity of the resulting composites is presented in Figure 3-15. Figure 3-15a illustrates the metallic behavior observed as function of Ni/ceria ratios when the composites

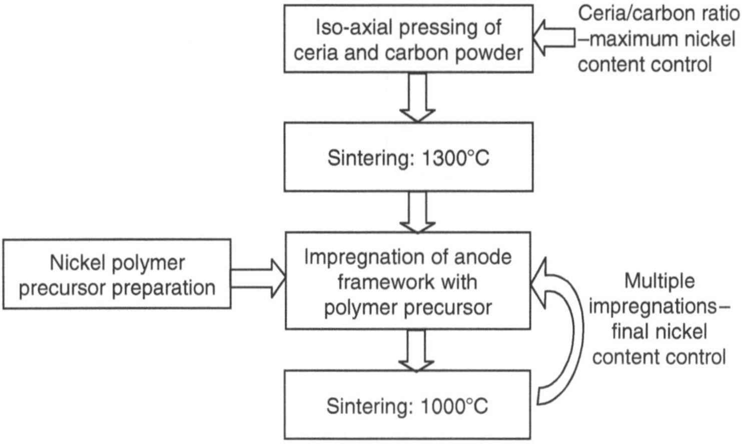

Figure 3-13. Schematic diagram of Ni–ceria cermet preparation

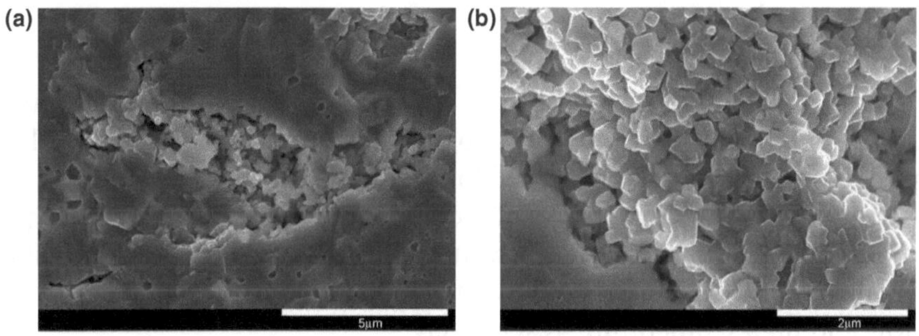

Figure 3-14. SEM pictures of the fracture of Ni–ceria cermet microstructure (11% of Ni): (**a**) ceria framework with nanocrystalline NiO; (**b**) NiO oxide area

were reduced in forming gas to form metallic Ni. The conductivity of the cermet with 14 vol% of Ni is slightly lower than that reported for the cermet with 35 vol% of Ni prepared by conventional techniques [17]. In the case of the cermet with 7.5 vol% of Ni the conductivity of about 10 S/cm is obtained, what is about two orders of magnitude higher than in the case of the cermet prepared by conventional techniques (0.1 S/cm) [17]. The conductivity of the cermet prepared by conventional techniques follows the percolation theory, which predicts a rapid increase of conductivity (percolation threshold) at about 30 vol% Ni [18]. This is not a case for net shape technology processed ceramics, which gives the opportunity to design materials with unique properties. The conductivity of the unreduced com-

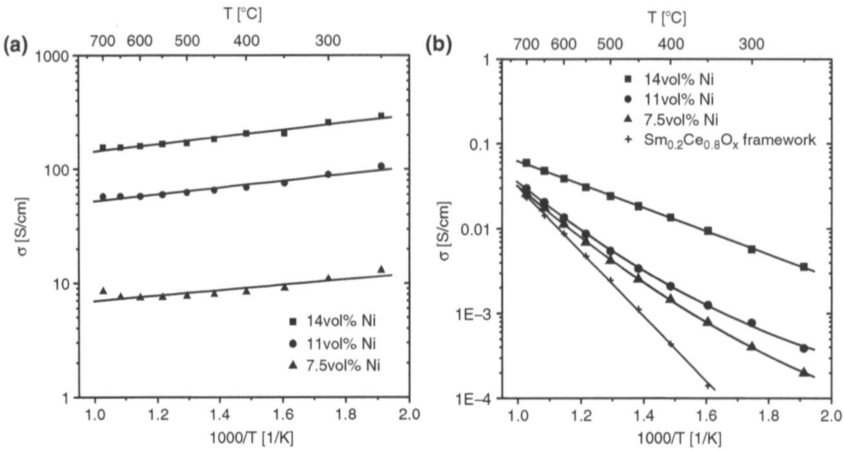

Figure 3-15. Temperature dependence of conductivity of reduced (**a**) and oxidized (**b**) nickel– ceria cermet

posites is shown in Figure 3-15b. The activation energy of the ceria framework (0.75 eV) and that of the 14 vol% of Ni composite (0.3 eV) reflects typical activation energies observed for ceria and Ni oxide, respectively. For the composites with 7.5 and 11 vol% of Ni the slope of the Arrhenius plot is different at low and high temperature. This is caused by the domination of Ni oxide and ceria conductivity in total conductivity of the composite at low and high temperature, respectively.

Anodes for fuel cell application should be stable both to redox cycling and to long-term operation in a reducing atmosphere. The electrical conductivity is a good indication of those features. Figure 3-16 shows the conductivity of a cermet with 11 vol% of Ni as a function of redox cycling at 700°C. As can be seen, the conductivity is not influenced by the redox cycling, which denotes that the cermet structure may be stable for long-term operation. This is related to the robust structure of the ceria framework, which was sintered at 1,300°C. Even the large volume changes which occur during oxidation/reduction cycling do not appear to harm the structure of the cermet. Long-term exposure of the cermet to a reducing atmosphere is known to cause Ni agglomeration (grain growth) and its dewetting from the oxide [19] which results in degradation of the electrical conductivity of the cermet. As can be observed in Figure 3-17, over the time of our experiments, the cermets with 14 vol% Ni appear to be stable to electrical degradation. Only the cermets containing lower Ni contents do show considerable degradation.

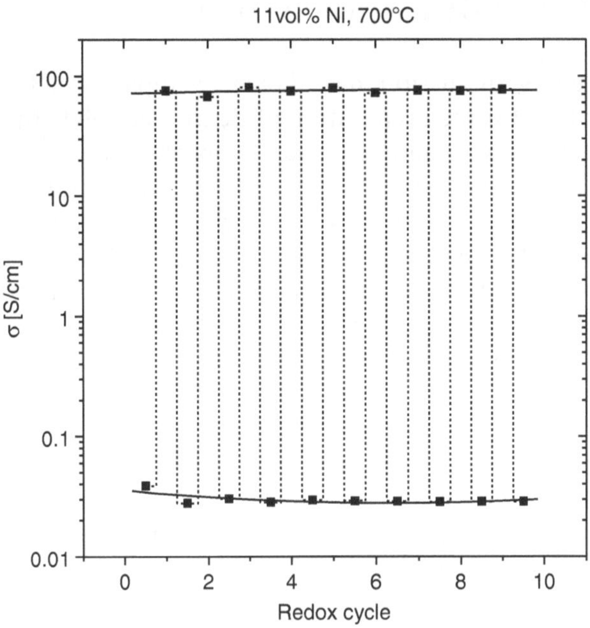

Figure 3-16. Stability of Ni–ceria cermet in redox cycling

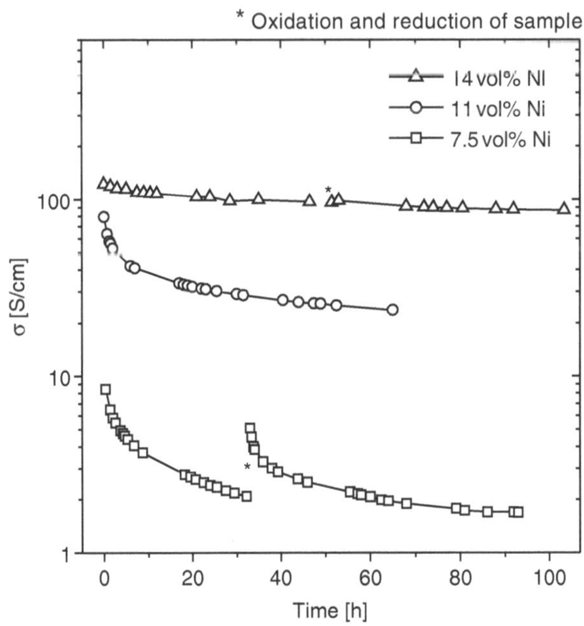

Figure 3-17. Stability of Ni–ceria cermet under reducing atmosphere

COMPOSITE ANODE AND CATHODE

The preceding discussion has shown examples which used the net shape technology for the preparation of dense electrolyte films, nanocrystalline bulk electrolyte, and electrode support. However, the net shape technology can also be successfully used for electrode film preparation. In this case this technology allows the tailoring of the ionic and electronic conductivity of the electrodes and their porosity. Below the examples of processing and properties of anode and cathode films will be presented.

As an example of anode the Ni–YSZ cermet was prepared and evaluated. A Ni–YSZ composite on sapphire was prepared according to the diagram shown in Figure 3-18 [20]. First, a 2–3 μm thick coating of YSZ colloidal particles were spin coated onto a sapphire substrate and then impregnated by YSZ polymer precursor. After each impregnation it was heated to 380°C to decompose organics. Next, the composite was backfilled by a Ni containing polymer precursor. After each deposition the composite was again heated to 380°C and after completion of the depositions it was annealed at 800°C. The samples included typically 30–40 wt% of Ni in the composite Ni–YSZ. Figure 3-19 shows SEM images of cross

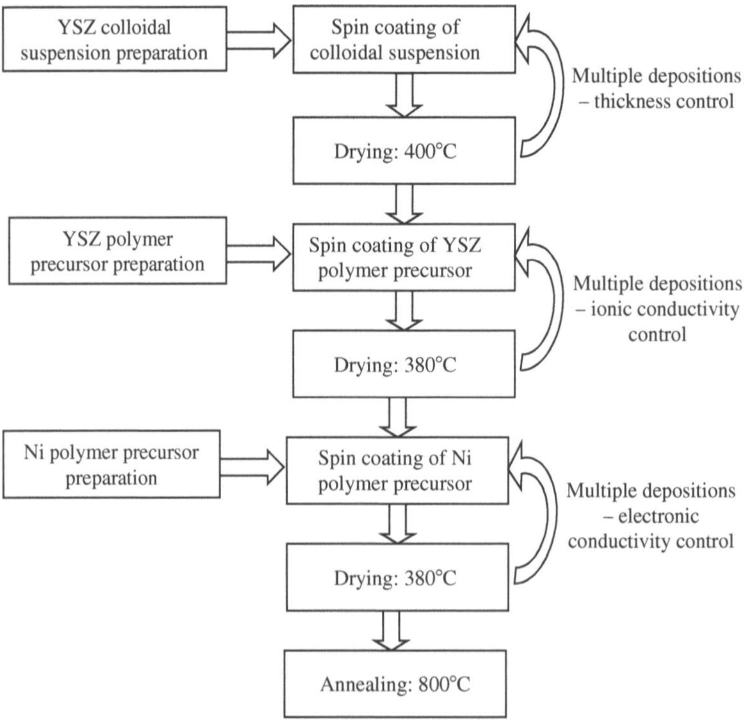

Figure 3-18. Schematic diagram of Ni–YSZ composite film preparation

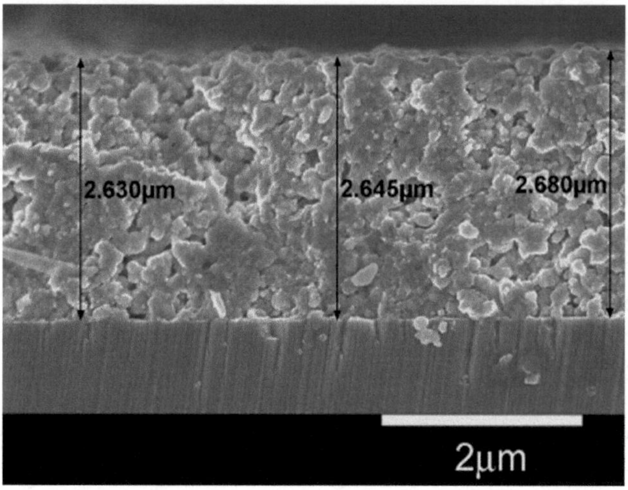

Figure 3-19. SEM image of the fracture cross section of the composite annealed at 800°C

sections of the composite Ni–YSZ on the sapphire substrate before reduction of the NiO. The thickness of the composite Ni–YSZ was about 2.6 µm. The conductivity of the composite Ni–YSZ for an initially oxidized sample in the reducing atmosphere, and an initially reduced sample in the oxidizing atmosphere both measured as the temperature increased from room temperature to 800°C is shown in Figure 3-20. As can be seen,

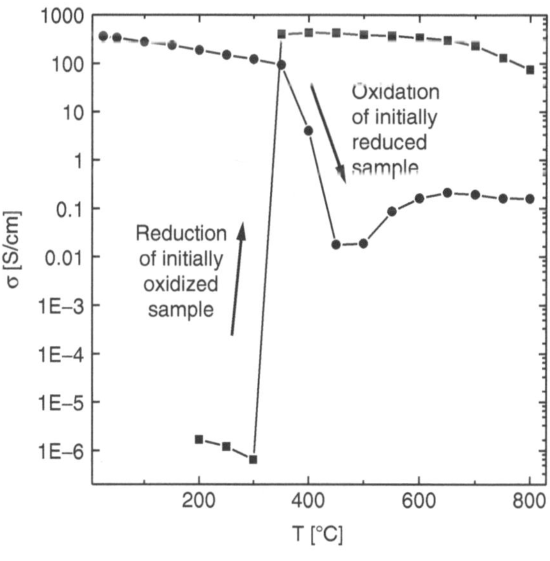

Figure 3-20. Electrical conductivity of the Ni–YSZ composite in the reducing and oxidizing atmosphere

the transition temperature from NiO to Ni (or Ni to NiO) of the composite Ni–YSZ appeared to be about 350°C. The conductivity of the composite Ni–YSZ was above 100 S/cm in the reducing atmosphere for all the temperature range, showing that it is suitable to use as an SOFC anode. The results of the oxidation and reduction cycling of the composite Ni–YSZ at 700°C are presented in Figure 3-21. The sample was cycled in reducing and oxidizing atmospheres with a 1 h hold for each. The conductivities in both reducing and oxidizing atmospheres were stable after the third cycle.

A life test results at 800°C in reducing atmosphere is shown in Figure 3-22. The conductivity was monitored for over 800 h. As can be seen the conductivity decreased to about half of the initial value during the first 50 h. However, after 50 h, the rate of conductivity decrease stabilized at about 0.02% h^{-1}. After 350 h, the sample was exposed to the oxidizing atmosphere for 1 h and then brought back into the reducing atmosphere several times. The stars in Figure 3-22 indicate when the oxidizing process took place. As can be seen, the conductivity in the reducing atmosphere increased after the oxidizing process to a value about 25% higher than it was prior to the life test and showed a decay rate of about 0.02% h^{-1} until the test was terminated at about 800 h. As the Ni particles were oxidized,

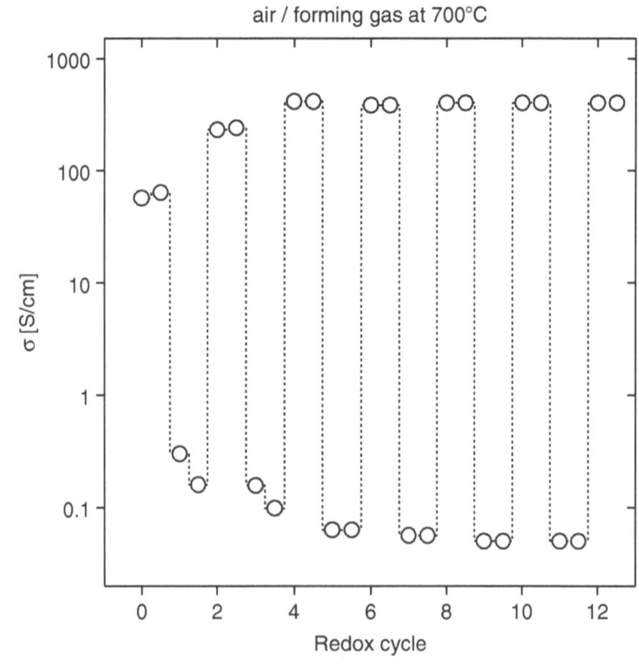

Figure 3-21. Stability of Ni–ceria composite film in redox cycling

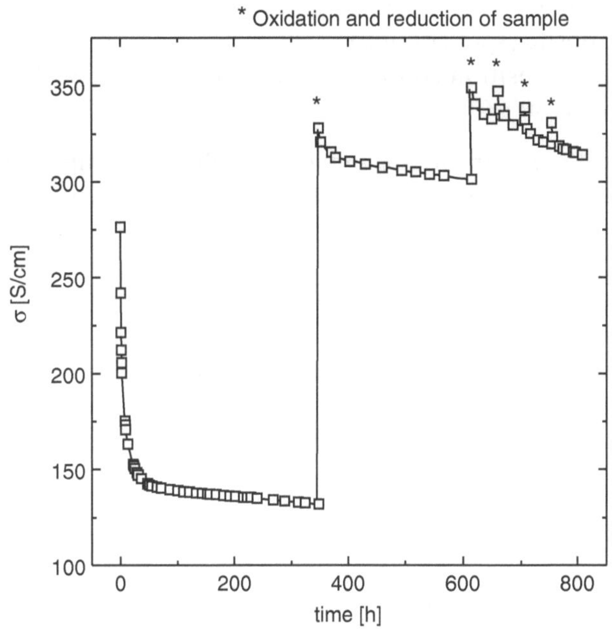

Figure 3-22. Stability of Ni–ceria composite film under reducing atmosphere

their volume expands and this may enhance contact between the NiO particles, which upon reduction may provide improved contact between the Ni particles.

As an example of a cathode film the $(La_{0.8}Sr_{0.2})_{0.9}MnO_3$ (LSM) perovskite will be used for electrode preparation and evaluation. Composite symmetrical and asymmetrical $(La_{0.8}Sr_{0.2})_{0.9}MnO_3$–YSZ (LSM–YSZ) structures were prepared on dense YSZ substrate (0.4 mm thick) [21]. The symmetrical cell was used for composite LSM–YSZ and composite LSM–YSZ/YSZ electrode overpotential evaluation, while asymmetrical for fuel cell performance and electrode overpotential measurement. The composite cathode was prepared similarly to the composite anode (Figure 3-18). The procedure follows:

- Preparation of YSZ colloidal suspension
- Preparation of polymer precursors (YSZ and target cathode materials)
- Coating the colloidal suspension on the substrate to prepare YSZ porous layer
- Depositions of YSZ polymer precursor to a porous YSZ layer to connect particles (spin coating)
- After each deposition composite was heated up to 380°C

- Depositions of LSM polymer precursor to provide a porous LSM–YSZ composite (spin coating)
- After each deposition composite was heated up to 380°C
- Annealing at 800°C

For the preparation of a symmetrical cell, the composite LSM–YSZ was prepared on both sides of a 0.4 mm thick YSZ substrate, which was used as an electrolyte. For a fuel cell preparation, prior to cathode deposition Ni–YSZ powder ink (Ni 45 wt%) was screen printed on the YSZ substrate and sintered at 1,400°C for 1 h.

The electrical conductivity of the composite LSM–YSZ as a function of reciprocal temperature is presented in Figure 3-23. The conductivities of YSZ [22] and LSM [23] were shown for comparison. The composite LSM–YSZ showed lower conductivity than bulk LSM simply due to lower concentration of LSM in the composite, however, the conductivity is sufficient for use as a cathode.

The area specific resistances for the composite LSM–YSZ symmetrical cell, which was obtained from impedance spectroscopic measurements, are shown in Figure 3-24. The R_2 and R_3 were attributed to electrode overpotentials (the charge transfer and the gas diffusion process), while the R_1

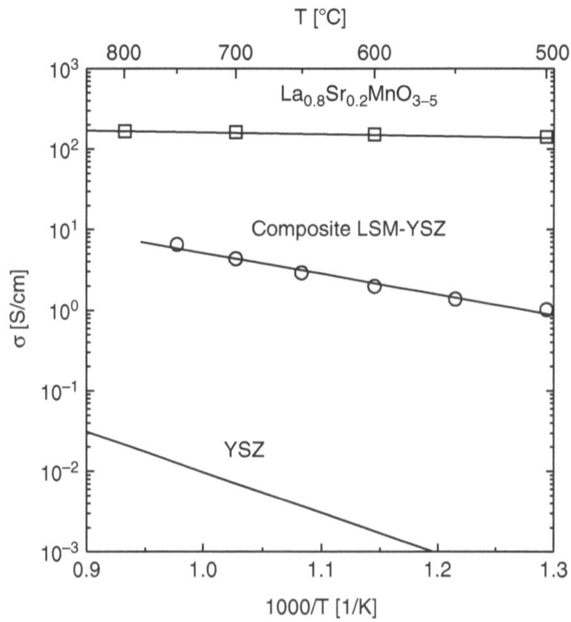

Figure 3-23. Electrical conductivity of composite LSM–YSZ along with the conductivity of bulk YSZ and LSM

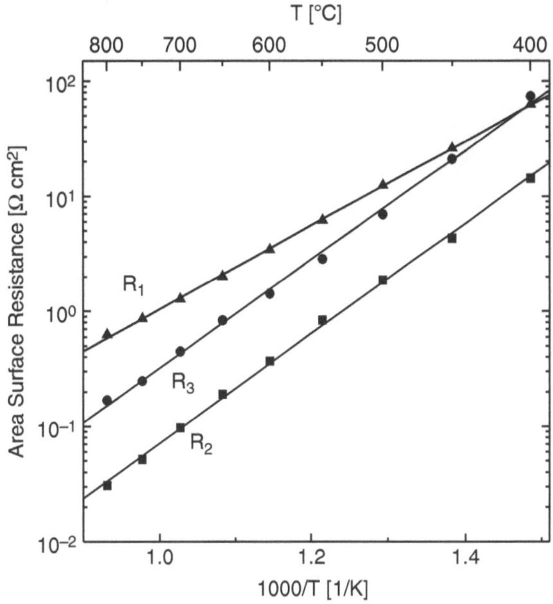

Figure 3-24. Area specific resistance of the composite LSM–YSZ symmetrical cell

to the electrolyte resistance. Overpotential resistances are low compared to the electrolyte resistance. The total overpotential resistance for the composite LSM–YSZ cathode was estimated to be 0.14 Ω cm^2 at 800°C and 0.3 Ω cm^2 at 700°C.

Fuel cell performance of the composite LSM–YSZ/YSZ/Ni–YSZ cell was investigated using forming gas (10 vol% H_2 in N_2) as the fuel (Figure 3-25). The results showed that a maximum power density of about 0.26 W cm^2 was obtained at a temperature of 850°C. The temperature dependence of the area specific resistances of the asymmetrical cell is shown in Figure 3-26. The electrode overpotential was estimated to 0.3 Ω cm^2 at 800°C, which is the total of anode and cathode overpotential. It appeared that about half of the overpotential originated from the anode, because the cathode overpotential determined from the symmetrical cell test was found to be about 0.14 Ω cm^2 at 800°C. The performance of the cell was mainly limited by the electrolyte resistance. The decrease in the electrolyte thickness would decrease electrolyte resistance. It can be concluded that the net shape technology can be successfully applied for the fabrication of cathode and anode electrodes.

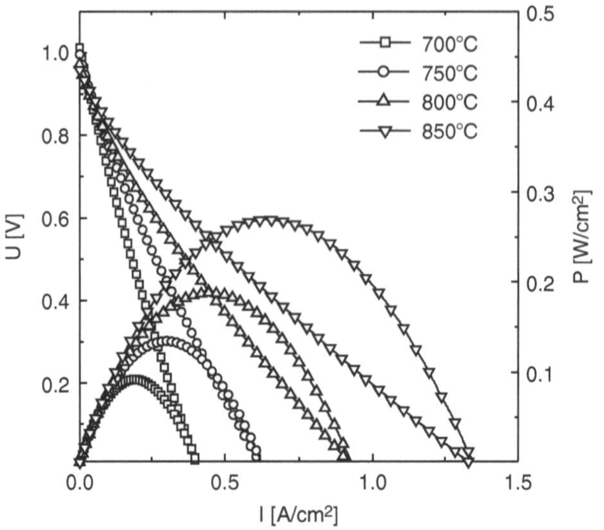

Figure 3-25. Electrical performance of the composite LSM–YSZ/YSZ/Ni–YSZ fuel cell

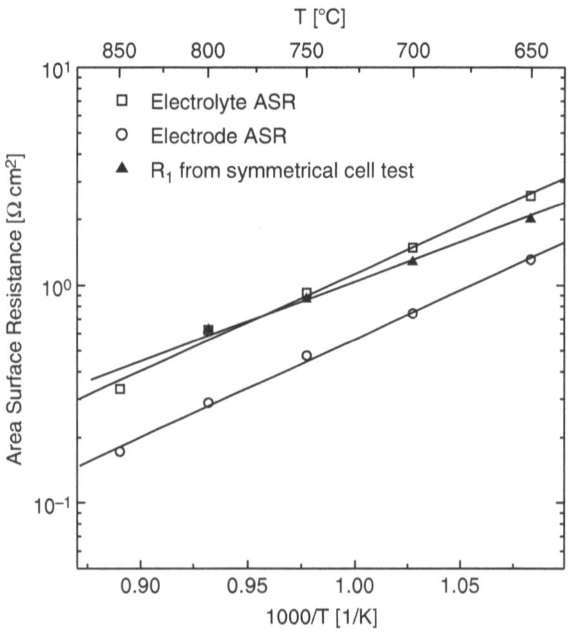

Figure 3-26. Area specific resistance of the composite LSM–YSZ/YSZ/Ni–YSZ fuel cell

CONCLUSIONS

In this text the possibility of application of novel technology, which is a combination of typical powder processing technology and polymer precursor method, for SOFC fabrication was presented. For examples of potential application in SOFC, the following structures were presented:

- Doped zirconia and ceria 1–10 μm thick electrolyte films
- Nanocrystalline bulk electrolyte support
- Ni-cermet anode support with low Ni content
- Ni-cermet anode films
- LSM cathode films

The following features which are attractive compared to current technology are:

- The possibility of preparation of dense 1–10 μm thick electrolyte films at temperatures below 400°C
- The possibility of preparation of the electrode layers with the required porosity and designed ionic and electronic conductivity
- The possibility of fabrication of required number of different layers using the same technology
- Percolation theory incompatibility that allows designing composites with special features
- Wide utilization due to cost effectiveness of equipment and operation

The understanding and exploiting all of these features requires further development of this technology. For example, the microstructure of the composite electrodes can be optimized or the composition of composite electrodes can be carefully selected in order to decrease the electrode overpotential resulted from the gas diffusion process or from the activation of electrochemical reactions, respectively.

References

[1] S.C. Singhal, *MRS Bulletin*, 25 (2000) 16–21.

[2] U. Pal and S.C. Singhal, *Journal of the Electrochemical Society*, 137 (1990) 2937.

[3] L.S. Wang, E.S. Thiele, and S.A. Barnett, *Solid State Ionics*, 52 (1992) 261–267.

[4] I. Kosacki, T. Suzuki, V. Petrovsky, and H.U. Anderson, *Solid State Ionics,* 136–137 (2000) 1225–1233.

[5] C. Wang, W.L. Worrell, S. Park, J.M. Vohs, and R.J. Gorte, *Journal of the Electrochemical Society*, 148 (2001) A 864–A868.

[6] V. Petrovsky, T. Suzuki, P. Jasinski, T. Petrovsky, and H.U. Anderson, *Electrochemical and Solid State Letters*, 7 (6) (2004) A137–A138.

[7] J. Will, A. Mitterdorfer, C. Kleinlogel, D. Perednis, and L.J. Gauckler, *Solid State Ionics,* 131 (2000) 79–96.
[8] P. Jasinski, V. Petrovsky, T Suzuki, T. Petrovsky, and H.U. Anderson, *Journal of the Electrochemical Society*, 152(2) (2005) A454–A458.
[9] M.C. Steil, F. Thevenot, and M. Kleitz, *Journal of the Electrochemical Society*, 144 (1997) 390–398.
[10] S.P.S. Badwal, *Solid State Ionics*, 52 (1992) 23–32.
[11] H.U. Anderson, M.M. Nasrallah, and C.C. Chen, US Patent 5,494,700, 1996.
[12] N. Oishi, A. Atkinson, N.P. Brandon, J.A. Kilner, and B.C.H. Steele, *Journal of American Society*, 88 (2005) 1394–1396.
[13] P. Jasinski, *Solid State Ionics*, 177 (2006) 2509–2512.
[14] X. Guo, W. Sigle, and J. Maier, *Journal of the American Ceramic Society*, 86 (2003) 77.
[15] J.A. Lane, J.L. Neff, and G.M. Christie, *Solid State Ionics*, 177 (2006) 19–25.
[16] P. Jasinski, T. Suzuki, V. Petrovsky, and H.U. Anderson, *Electrochemical and Solid State Letters*, 8(4) (2005) A219–A221.
[17] S.P. Jiang and S.H. Chan, *Journal of Materials Science*, 39 (2004) 4405–4439.
[18] N.Q. Ming and T. Takahashi, Science and Technology of Ceramic Fuel Cells, Elsevier Science B.V., Amsterdam, 1995, p.156.
[19] S.P.S. Badwal, K. Foger, *Ceramics International*, 22 (1996) 257–265.
[20] T. Suzuki, V. Petrovsky, P. Jasinski, and H.U. Anderson, *Electrochemical and Solid State Letters*, 8(7) (2005) A341–A343.
[21] T. Suzuki, M. Awano, P. Jasinski, V Petrovsky, and H.U. Anderson, *Solid State Ionics*, 177 (2006) 2071–2074.
[22] T.H. Etsell and S.N. Flengas, *Chemical Reviews*, 70 (1970) 339.
[23] J.H. Kuo, H.U. Anderson, and D.M. Sparlin, *Journal of Solid State Chemistry*, 87 (1990) 55.

Chapter 4

IN SITU SEAL INTEGRITY SENSING FOR SOLID OXIDE FUEL CELLS

John Olenick, Dr. Viswanathan Venkateswaran,
Tim Curry,[1] Robert Bourdelaise, Eli Richards,
Paul Vichot, and Barry Grabow[2]
[1]ENrG Incorporated Buffalo, NY, USA, [2]Johns Hopkins University Applied Physics Laboratory (JHUAPL), Baltimore, MD, USA

INTRODUCTION

Solid oxide fuel cells (SOFCs) involve structures composed of multiple ceramic layers, or interleaved layers of ceramic membranes and metal interconnects. To eliminate premature mixing of fuel and air gases or leaking of these gases from interior regions of the structure, the interfaces of adjacent layers are sealed with a glass or ceramic seal. These seals must withstand the high-temperature environment of the SOFC over its lifetime. Therefore these materials must be thermally matched with the adjacent layers to minimize transient stress risers and eliminate the potential for consequent seal failure.

The conventional approach used when designing and manufacturing seals for high-temperature applications is to develop mathematical models for calculating the seal stresses that are likely to arise during device operation, and thereby attempt to predict operational regions of safety or failure. This method always requires correlation to the modeled effects for clarity in the design process. It is inherently difficult to monitor seal integrity once an SOFC has been fabricated. Further, failures are typically catastrophic and destroy the evidence of the cause. Any attempt to disassemble a failed device may also destroy the evidence for failure.

In situ stress sensing methods relating real time results would permit operational performance stresses to be correlated to input variables thereby permitting the creation of a safe operational zone. Direct feedback to a closed-loop control system can then constrain the device within the safe zone. An in situ seal integrity sensing design and methodology has been demonstrated by ENrG Inc. and JHUAPL.

HIGH TEMPERATURE ISSUES

Typical planar SOFC stacks are composed of various ceramic materials that exhibit a range of high thermal expansion coefficients, low thermal shock resistance, low fracture toughness, and poor mechanical strength characteristics. Stacks are layered structures composed of interleaved cells and either metal or ceramic interconnects. The interconnects are perimeter sealed to the cells to eliminate the possibility of air and fuel prematurely mixing around the cell instead of through the ceramic membrane of the cell component. Various ceramics are used to produce the cell component, which is a thin membrane of an ionic conductor with an anode and cathode on either side. All of these functional layers are either ceramic or cermet materials capable of handling reducing conditions on the anode side and oxidizing conditions on the cathode side. At temperatures greater than 500°C, oxygen ions readily conduct from the cathode side of the ceramic membrane to the anode side where they combine with hydrogen ions creating water, heat, and electricity. As long as fuel and air are flowing to the stack, it will become thermally self-sustaining.

For SOFC systems operating at high temperatures (>650°C), the thermal expansion coefficients of the cell components, interconnect structures, and hardware materials must be carefully matched to prevent the development of severe thermal stresses resulting in cell or seal failure during various operational modes. Although mathematical models have been used to calculate the generated stresses in the stack and attempt to predict operational regions of safety or failure, the complexity of a fuel cell stack makes it very difficult to quantify stresses that develop during operation, until of course a catastrophic failure occurs. Unfortunately with thin membrane ceramics or seals, these failures can be so catastrophic that if the failure itself does not destroy the ability to evaluate the very reason for the failure, the disassembly of the stack may result in the loss of the data. Therefore, other in situ evaluation and ultimately control methodologies are necessary to further the development, commercialization, and reliability of SOFC-based energy systems.

TDR AND FR

Ultimately, coupling high temperature in situ evaluation techniques with correlation to stress models and predictive equations can result in the ability to produce a closed-loop in situ control system that is cell and seal sensitive and permits safe stack operation in all modes while minimizing the resultant stress modes. To achieve such a control system requires the incorporation of seal structures which can be electronically tested using time domain reflectometry (TDR) and frequency response (FR). These methods can appraise the integrity of the cell membrane and perimeter seal through analysis of the structure's electronic "signature" in response to pulsed and microwave signals. The incorporation of multilayer signal structures in the cell membrane and ceramic/glass seal tied to a signal generator and analyzer yields structural radar type images showing changes in cross section, build up or relaxation of stress, existence of discontinuities, or ultimately failure of the ceramic medium.

The use of TDR and FR signal methods to extract an electronic "signature" from fiber optic cables, printed circuit boards, moisture in soils measurements are well known. These techniques have yet to be applied to the field of energy production and specifically to SOFC stacks. Both techniques are classic methods for diagnosing failures along interconnects of the "controlled impedance" variety, such as high speed digital circuit boards, and to recognize small changes in performance in light of minor physical changes, such as those found in RF assemblies. TDR and FR techniques have been used in multiple cases to discern thermal stress anomalies in ceramics. Anderson [1] used TDR on difficult to fabricate high temperature superconducting multilayers. Anderson was able to fabricate a simple circuit structure in the multilayer leading to the full characterization of the high-speed digital waveform and pulse integrity. Additionally, Anderson performed thermal cycling on the structures to identify if any thermal expansion mismatches occurred. Swaminathan [2] used high-speed analysis techniques to analyze process defects in embedded passive components such as capacitors and resistors in ceramic multilayers.

Demonstration of the creation of multilayer line structures in the seal will result in the collection of physical measurement data on the change in seal features during heat up, steady state operation, transients, or on cool down. A radar type image of the signal characteristics will show build up or relaxation of stresses, discontinuities in seal structure including porosity, delamination of the seal from either the cell or interconnect, and ultimately the catastrophic failure of the seal. Additionally with the ability to use a metal interconnect as one of the plane references coupled with a trace on the oxide coating on the interconnect, electrical conductivity and structural integrity of the thin oxide film on the metal interconnect can be determined within the vicinity of the trace.

SEAL DESIGN

The seal design includes an electrical transmission line embedded within dielectric material sealing an interface between adjacent structures. A signal injection port is provided for exciting the transmission line with a signal source and a second port allows for sampling the transmission line while the excitation signal is present. The sample ports were adapted for connection to test and measurement apparatus which are used to analyze signal samples for indications of seal integrity problems. Using TDR or FR analysis, the transmission line was monitored for changes in characteristic impedance due to variation in seal dielectric constant and/or disruption of the transmission line. Such an evaluation provided useful information for making determinations about seal integrity in a manner that permitted safe seal operation and minimization of resultant stress modes.

Creating a seal in the image of a passive electronic component may be accomplished by building up the seal in a multilayered approach as shown in Figures 4-1 and 4-2. The perfect seal maintains a consistent and expected signal response over temperature. Anomalous behavior would give a characteristic signal response. Anomalies of particular interest are breaks in the seal, effacing of the seal wall, and changes in seal dimension due to thermal stress.

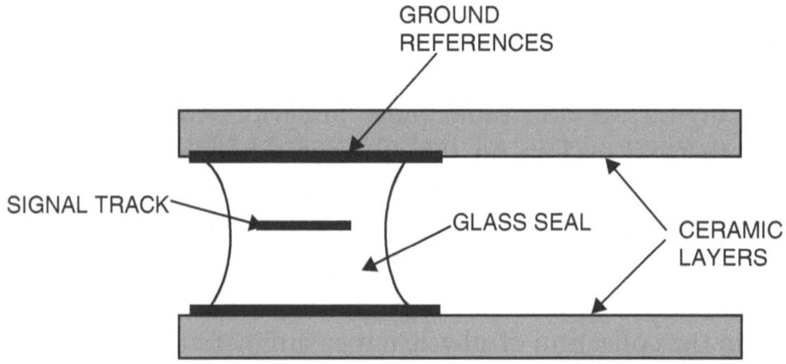

Figure 4-1. Construction of a seal that also functions as a transmission line interconnect

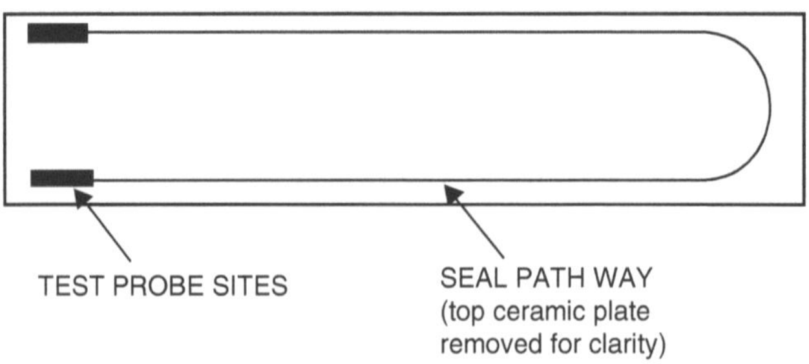

Figure 4-2. Plan view of a 2 in. × 6 in. seal test sample monitored using TDR and FR methods

Inserting a pulse (TDR) that rises to final value within picosecond durations and listening to the "return" pulse is similar in principle to a radar or sonar system. The method can detect anomalies and their location along the line, but may not have the resolution to detect the seal's proximity to failure. In contrast, inserting a spectrum of sinusoids (FR) into the seal and measuring the energy that is transmitted through or reflected back at each frequency can detect minor changes in seal integrity and proximity to failure but not location of an anomaly along the line. In effect both techniques must be used in this phase of the program in order to validate the type and location of the seal anomaly with its corresponding "electronic signature."

There are two risks associated with this approach: First, the ability to create a seal with the required dimensions of the signal track and associated dielectric constant of the glass to yield sufficiently high impedance (target 50 ohms) and second, that the anomalies have recognizable and reproducible signatures within the passable bandwidth of the interconnect.

The TDR method measures the characteristic impedance along a transmission line. To be effective, the transmission line must be contiguous along the length of the seal and may turn corners. Ideally the seal must be accessible at both ends. Figure 4-1 illustrates a longitudinal cross section of a seal with an embedded transmission line and the instrumentation probe feature where the signal is inserted or observed. The geometry of the signal insertion/probe point must be outside the cell seal area and is fabricated by the same screen printing process that creates the seal in a multilayer fashion. Figure 4-2 illustrates the plan view of a possible test sample (top ceramic layer removed for clarity).

In operation the ideal response from the TDR signal will show a flat response at the impedance value of the line. Corners, breaks, deviations in impedance due to stress, and their distance from the signal source should all recognizable features. Figure 4-3 illustrates a possible TDR response with several different kinds of seal anomalies.

FREQUENCY RESPONSE APPLIED TO SEAL INTEGRITY

The TDR method sends a "step function" (containing a broad spectrum of signals) down the seal and observes the response in the time domain. In contrast, the FR method sends a narrower spectrum of sinusoids down the seal and observes the response in the frequency domain by comparing the signal energy sent at each frequency with the energy making it through the line (S_{21} or forward gain), and that reflected at the source (S_{11} or return loss). Figure 4-4 is an illustration of a notional S_{21}

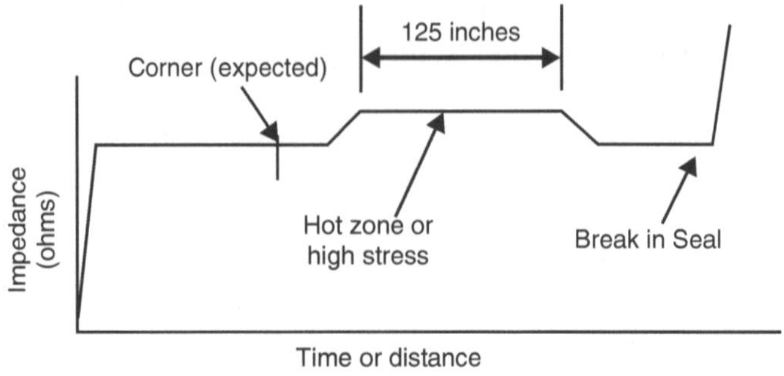

Figure 4-3. Notional TDR response of a seal with anomalies

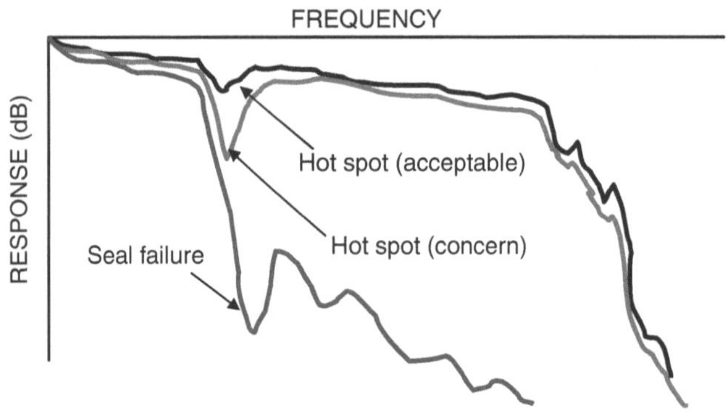

Figure 4-4. Notional S_{21} responses superimposed of a seal with a hot spot identified, growth in severity and final failure

response of a hot spot growing in severity until failure. It is envisioned that the FR technique could be used to recognize anomalies and their severity. The FR method views the seal as a network made up of passive components (such as a capacitors, resistor, or inductors), which have a unique response "signature." Changes in seal integrity, as mentioned above, will all be observed as characteristic differences at unique frequency or band of frequencies. Although the location of the anomaly cannot be determined, the resolution and awareness of the proximity to failure should be superior to the TDR method. Both techniques are necessary, however, to establish and validate a link between a particular anomaly and its impact on the FR signature.

EVALUATION TESTING

It was apparent that voids were created in the top/bottom of the dielectric seal during processing of the samples that contained the microstrip transmission line. For this reason, a majority of the testing was performed on samples containing the coplanar transmission line.

The fuel cell seal samples were tested using the Microtech Model 44 Probe Station. A block diagram of the TDR measurement system is shown in Figure 4-5. The sample is mounted on the wafer chuck, and both ends of the transmission line embedded in the sample are connected to the TDR sampling head.

A block diagram of the network analysis measurement system is shown in Figure 4-6.

A propane torch was used to heat selected samples. The samples were cooled using compressed air.

RESULTS

The variation in the impedance of the microstrip transmission line samples was approximately 150% greater than that of the samples that employed the coplanar type transmission line. Figure 4-7 shows a composite of the coplanar waveforms while Figure 4-8 shows a composite of the microstrip waveforms.

TDR measurements of a sample subjected to localized heating and cooling showed a variation of approximately 5% from a recorded baseline measurement of the reflected signal in the region at which the temperature source was applied.

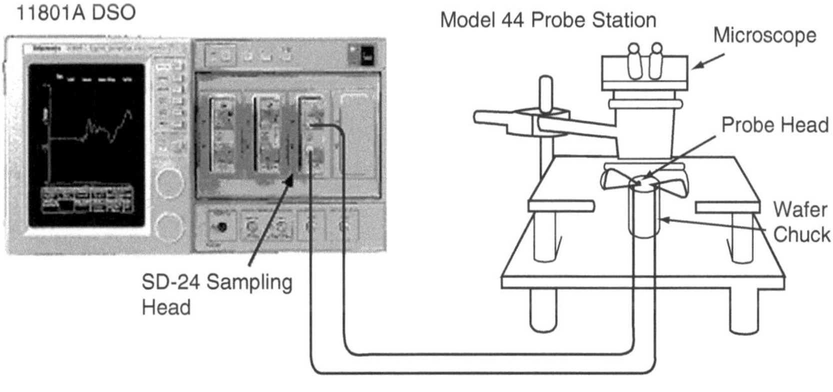

Figure 4-5. Block diagram of TDR measurement setup

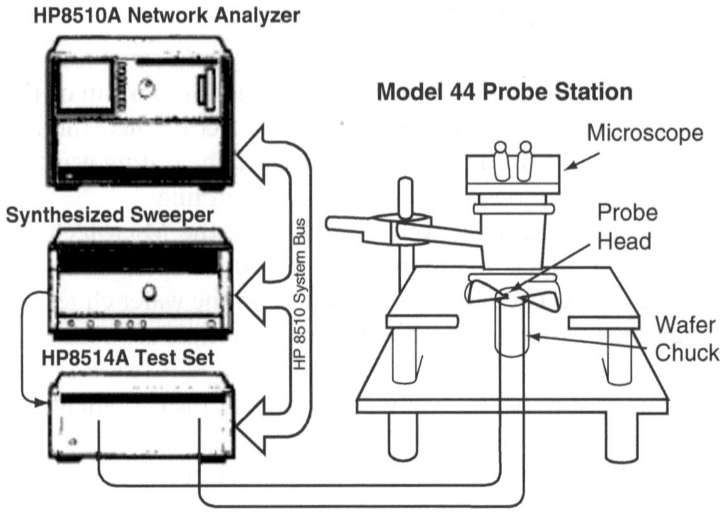

Figure 4-6. Block diagram of network analyzer test setup

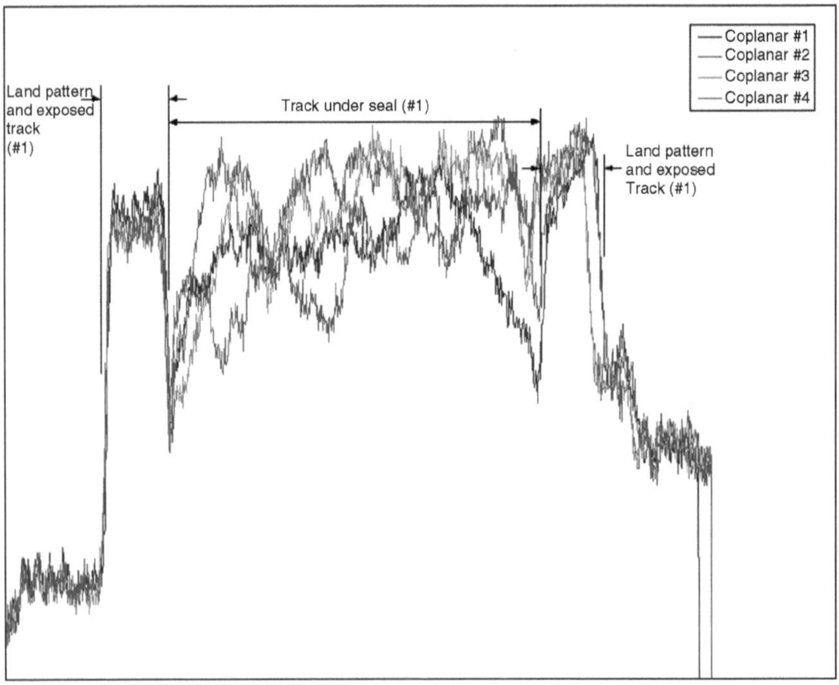

Figure 4-7. Composite of TDR waveforms from four coplanar waveguide samples

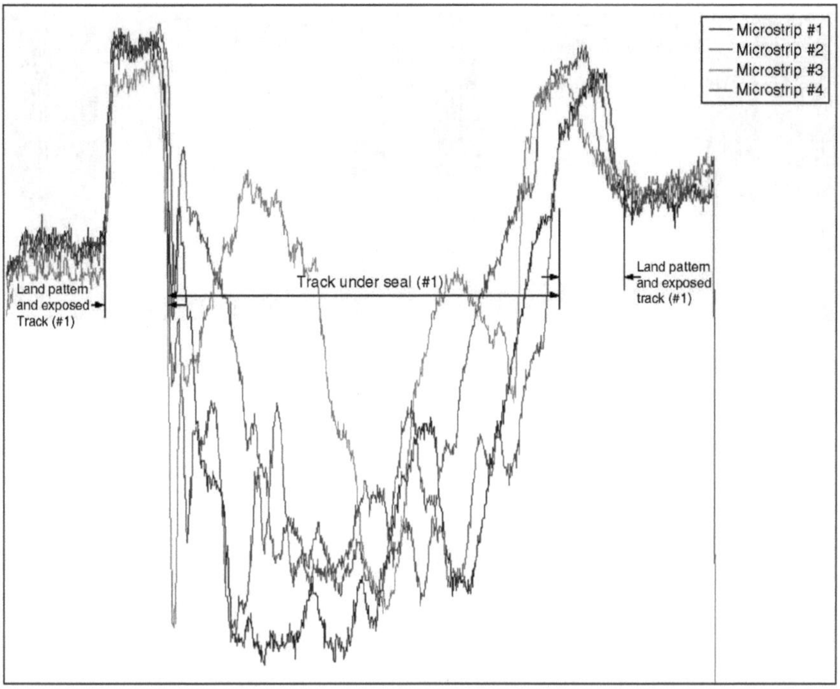

Figure 4-8. Composite of TDR waveforms from four microstrip samples

Baseline TDR measurements were made for two samples in their entirety and then the two samples were ablated. In one case a laser was used to make a cut perpendicular to the direction of the track formed by the dielectric seal material. In the second sample, a dicing saw was used to cut through the top plate of the sample and remove a portion of the dielectric seal. After the samples were ablated, a second set of TDR measurements were made for comparison with the baseline. In both cases, there were no visible differences between the baseline and the second measurement.

In a third sample, two dicing saw cuts were made perpendicular to the track direction and parallel to one another as shown in Figure 4-9, allowing a section of the ceramic top plate and a print of the dielectric seal material to be removed. In this sample, TDR measurements comparing the same sample in cases both with the ceramic "lid" in place and with the lid removed showed a visible phase-shift in the time-domain signal.

Network analyzer measurements showed very frequency-selective changes in reflected and transmitted power in response to localized changes in the dielectric constant or to heating of the sample. The measurements were made over the range between 45 MHz and 26.5 GHz. In most cases,

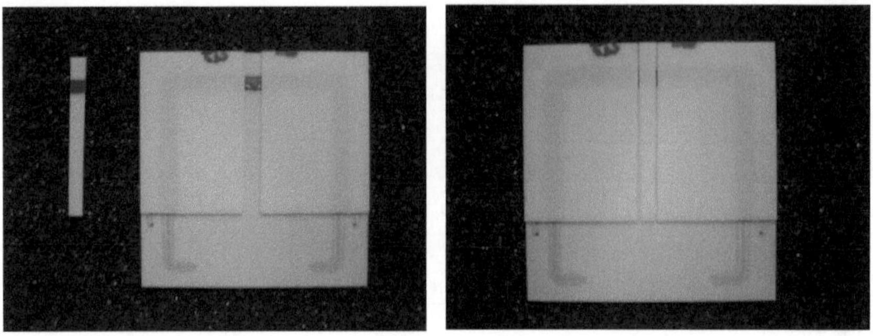

Figure 4-9. Coplanar sample #5: ceramic cutout shown removed from sample (left) and placed on sample (right)

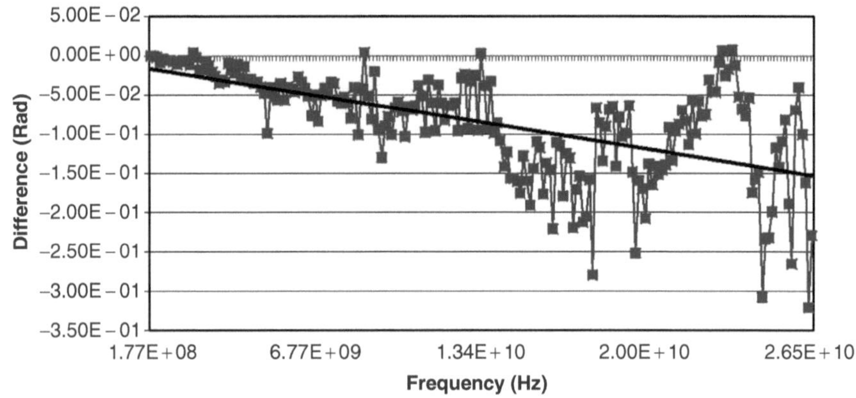

Figure 4-10. Plot of S_{21} phase difference. Difference calculated for two sample #4 measurements at cold and hot temperatures (line is first order trendline)

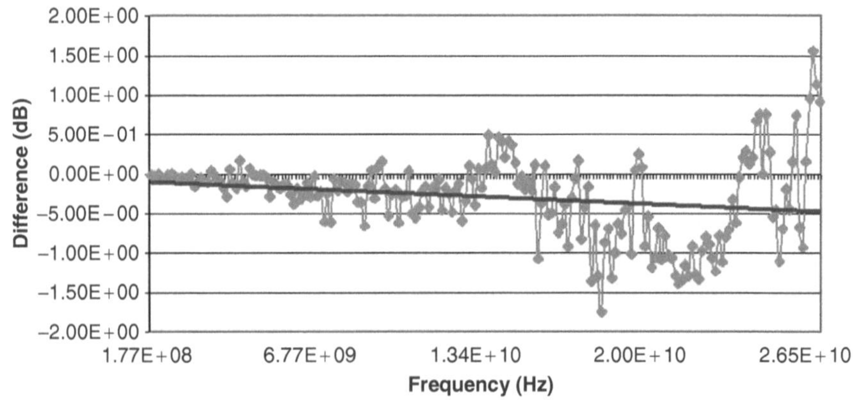

Figure 4-11. Plot of S_{21} magnitude difference. Difference calculated for two sample #4 measurements at cold and hot temperatures (line is first order trendline)

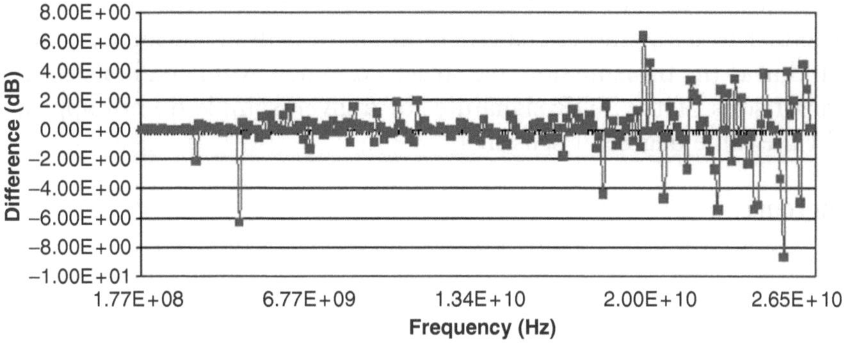

Figure 4-12. Plot of S_{22} magnitude difference. Difference calculated for two sample #5 measurements with and without "lid"

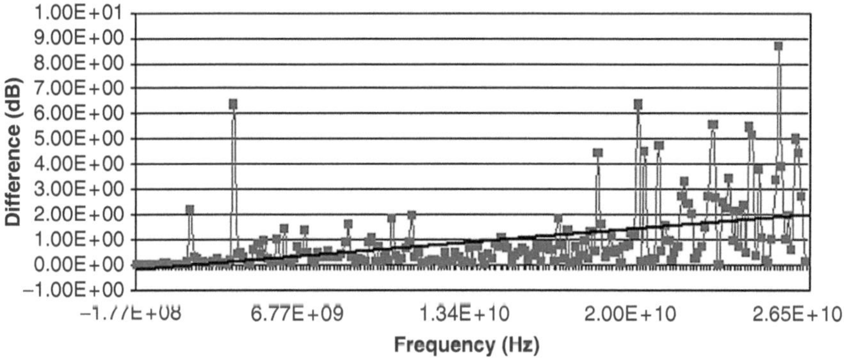

Figure 4-13. This plot shows the absolute value of the difference calculation in Figure 4-12, illustrating increasing variability in magnitude as frequency increases (line is first order trendline)

the largest changes occurred in the high end of this spectrum. These results are shown in Figures 4-10–4-13.

The test results suggest that the fuel cell seal variant which contains the coplanar waveguide structure is probably a better candidate than the microstrip variant for the goal of designing a seal that might be indicative of defects or precursors to seal failure, as the sample-to-sample variability is much greater in the microstrip variant.

CONCLUSIONS

A comparison of the TDR and the FR methods for fault detection/measurement suggests that high-frequency measurement of transmitted/reflected power (S_{21}/S_{11}) and phase using the FR method is the most effective means of detecting changes in the seal characteristics due either to temperature changes or to changes in the dielectric constant in and near the transmission line structure. When evaluated in its entirety, comparisons of the TDR signatures of a given sample do not appear to offer the same degree of sensitivity.

References

[1] P.M. Anderson, A.W. Lindner, P.M. Chau, and A.D. Smith, High Temperature Superconducting Multilayer Multichip Module: Fabrication and High Speed Characterization, Manuscript on work supported by Navy SBIR Contract #N61331-96-C-0017 at http://www.st.northropgrumman.com/whoweare/SiteFiles/docs/stratedge paper.pdf
[2] M. Swaminathan and S. Pannala, Mixed Signal Portable Wireless Packaging Design and Test Challenges, *Proceedings of the IEEE Computer Society Workshop* on VLSI '98, Orlando, Florida, April 1998, pp. 89–94.

Chapter 5

SOLID OXIDE FUEL CELL

Dr. Xuan Wang
Intel, Sr. Packaging Engineer, 5000 W. Chandler Blvd., Chandler, AZ 85226, USA

INTRODUCTION

Fuel cell is the electrochemical device which converts chemical energy to electricity. Unlike the internal combustion power generators, which convert chemical energy to electrical energy through an intermediate step-mechanical energy, fuel cells directly convert chemical energy to electricity. Therefore, the efficiency is much higher than the combustion power generator.

The first solid oxide fuel cells (SOFCs) emerged in 1930s by Emil Baur and H. Preis. The materials they used are zirconium, yttrium, cerium, lanthanum, and tungsten oxide [1]. SOFCs differ from other types of fuel cells in that the components of the cell materials are solid instead of liquid; especially the electrolyte is solid oxide, usually ceramics due to the high working temperature. Extensive efforts to develop SOFC for transportation, the bottoming cycle of a power plant, and distributed generation of electric energy are motivated by a need for greater fuel efficiency and reduced air pollution. SOFCs are generally ideal for stationary high power applications with an output from 1 kW to 2 MW. Their working temperatures are typically between 700 and 1,000°C, which is much higher than other types of fuel cells. The large amount of heat generated by SOFC is usually utilized to drive secondary gas turbine in order to increase the efficiency. This type of hybrid system is expected to reach the efficiency up to 70% [2]. The fuels to power SOFC can be hydrogen and other hydrocarbon fuels, such as natural gas and carbon monoxide. Air (or oxygen) is used as oxidant [3]. This is very important due to the handling and storage issue of hydrogen. The existing infrastructure of hydrocarbon fuels can be easily available for SOFC. The only by-product of SOFC is water and heat, which has no detrimental impact to the environment.

The unique working temperature and the structure of SOFC give it the following advantages:

1. High efficiency compared direct combustion power generator.
2. Suitable for hydrogen fuels as well as hydrocarbon fuels, makes it compatible with current energy infrastructure. Hydrogen provides a potential clean energy solution for large-scale deployment.

3. Environment friendly: unlike batteries usually containing elements harmful to health, the materials in SOFC have less impact to the environment.
4. There is no liquid electrolyte involved, so it is easy to operate and maintain.
5. The high operating temperature avoided expensive catalysts.
6. The working temperature of SOFC is high enough to allow for internal reforming, which means to produce hydrogen from different types of fuel gases within the cell stack itself (without external reformer).
7. There are no moving parts, so no vibration and less noise.
8. SOFC can be operated as auxiliary power unit to utilize the heat generated, and further increase the overall efficiency.
9. Suitable for large, high power applications, such as power generation stations.

There are also some disadvantages which prohibit the large-scale deployment of SOFC:

1. Due to high operation temperature, the life of SOFC components is limited.
2. High manufacturing cost of components, especially for the interconnection materials.
3. Coefficient of thermal expansion (CTE) mismatch during thermal cycling causes stresses; the stresses cause delamination at interfaces.

STRUCTURE

The SOFC unit has a three-layer sandwich structure: two porous electrode, anode and cathode, serving as the chemical reaction, and the electrolyte, serving as the diffusion layer of oxygen ions but electrically nonconductive. A typical SOFC structure is shown in Figure 5-1. The anode and cathode should be porous to allow the diffusion of oxygen ions. A single SOFC unit cannot provide enough power; therefore, the interconnection between stacks of cells is required.

At the cathode, oxygen or air is reduced to oxygen ions:

$$\frac{1}{2}O_2 + 2e^- \rightarrow O^{2-}$$

The cathode has to be porous to allow oxygen to flow through it to the electrolyte. It also must have low resistivity to conduct electrons. It is required that the cathode material has excellent electronic conductivity as well as thermal stability. Therefore, noble metals and electronic conductive

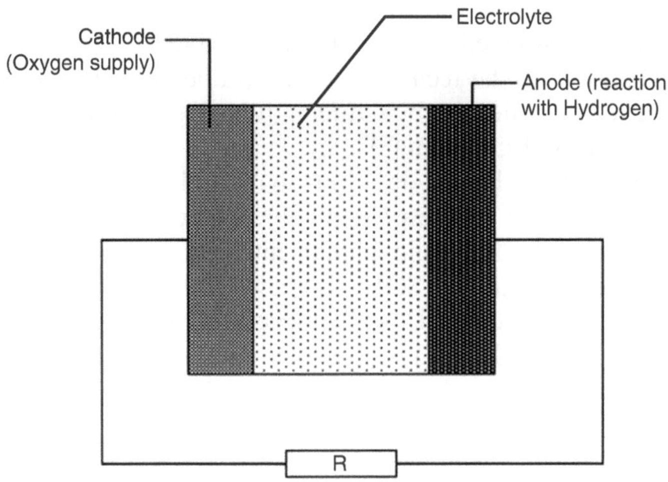

Figure 5-1. Structure of SOFC

oxides are used as cathode materials. The cathode must be porous so that it permits rapid mass transport of reactant and product gases.

At the anode, the oxidation reaction occurs when fuel reacts with the oxygen ions to generate electricity and water:

$$O^{2-} + H_2 \rightarrow H_2O + 2e^-$$

The function of electrolyte is to separate cathode and anode electronically so the electrons are "forced" to flow through external circuit. At the same time, the electrolyte must allow the oxygen ions to diffuse through it. The electrolyte usually determines the operating temperature of the fuel cell. It also allows the flow of charged ions from one electrode to the other to maintain the overall electrical charge balance.

From the mechanical support point of view, SOFCs can be divided into two main groups: the electrolyte-supported and electrode-supported SOFC. Usually the support layer is the thickest to provide enough mechanical strength for the cell.

From the structure point of view, there are two types of structures of SOFC: tubular and planar. Tubular SOFCs have shown some desirable characteristics over systems with planar SOFCs [4]. Tubular SOFCs can alleviate the sealing problem arose by CTE mismatch of planar SOFC; therefore, they are robust for repeated cycling under rapid changes in electrical load and in cell operating temperatures. The large form factor tubular SOFC built by Siemens Westinghouse has successfully conducted long-term operation over 70,000 h. Small-scale tubular SOFCs could endure thermal stress caused by rapid heating up to operating temperature. Sammes investigated anode-supported tubular SOFCs and started commercialization of the technology. Microtubular SOFCs can increase the volumetric power density by reducing tube diameter, which is not possible with planar SOFC design [8].

In planar SOFC, the anode, electrolyte, and the cathode are sandwiched together; their geometry is planar. The cells were assembled by stacking planar cells and rigid planar interconnectors alternately. The typical structure of a tabular SOFC is shown in Figure 5-2.

The planar structure SOFC is a composite structure which consists of heterogeneous materials. CTE mismatch causes warpage and residual stresses during heating and cooling of the SOFC. The residual stresses causes the voids and delamination between layers and eventually decreases the electrical conductivity between SOFC cells. The curvature effects oppose successful stack operation, which also relies on the permanent physical contact of geometrically stable planar cells. Cell flatness can be achieved with a compensation layer but the average residual tensile stress in the anode increases. On the other hand finite element analysis (FEA) indicates that relatively high loads are necessary to completely suppress cell curvature [5]. FEA modeling results indicated that the thermal stresses in the three layers of the cell are mainly due to the heating from room temperature to the operating one [6].

Other new type of structures have been developed to achieve a tight gas seals and sufficient electrical contacts between each cell. In the envelope

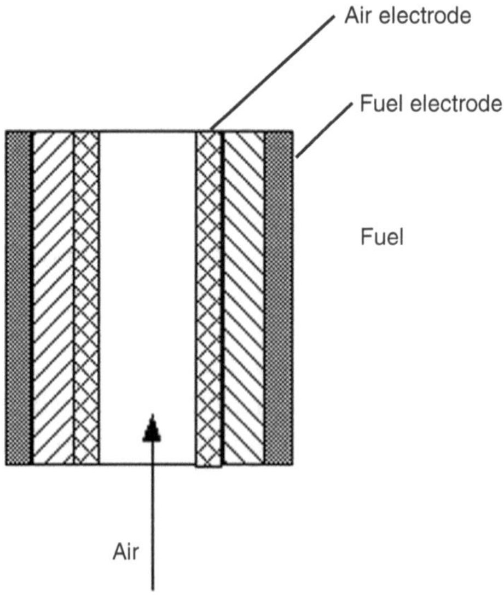

Figure 5-2. Structure of tubular SOFC

structure developed in Japan, each cell is wrapped in a thin electrically conductive alloy and connected to the next cell through a flexible metal interconnector. Tight gas seals and sufficient electrical contact between each cell are achieved. Seven- and twenty-cell stacks have been tested and the result is promising [7].

The single tubular cell with 1.6-mm diameter and 1-cm length, generated about 140, 300, and 350 mW at 500, 550, and 570°C with H_2 fuel, has been reported. This indicates that a cell stack of 55-cell arrangement whose volume is about 1 cm can possibly generate 3.5, 7.5, and 8.7 W at 500, 550, and 570°C, respectively, using the microtubular cells [8].

Porous Ni anode has also been employed in tubular cells and proved stable. A porous nickel tubular support of 1-mm diameter was successfully created by heat-treating a commercial nickel tube. This is suitable candidate for microfuel cells [9].

Due to the high operating temperature and thermal cycling of SOFC, degradation, residual stresses caused by CTE mismatch, warpage of stacked planar cells, and the reliability of the sealing are all challenging problems. The manufacturing of these materials is usually costly, which prevents SOFC commercialization. Many researches have been investigated to resolve these problems from materials, cell structure, lowering the operating temperature, and new design. Low operating temperature enables SOFC to use less expensive metallic separators for cell stacking and to carry out internal reforming of hydrocarbon fuels.

Reactive air brazing method has been investigated for use in sealing various high-temperature solid-state electrochemical devices, including planar SOFCs. The tests using a filler metal composed of 4 mol% CuO in silver showed no change in rupture strength or loss in hermeticity after 800 h of high-temperature exposure, but did undergo microstructural change due to the dissolution of hydrogen into the silver-based braze material. Air-brazed specimens subjected to rapid thermal cycling exhibited no loss in joint strength or hermeticity, but displayed initial signs of seal delamination along the braze–electrolyte interface after 50 cycles [10].

CATHODE

Sr-doped $LaMnO_3$ (LSM) cathode have been extensively investigated and developed as electrode materials. For oxygen reduction in SOFCs, mixed ionic and electronic conducting (MIEC) materials such as (La, Sr)(Co, Fe)O_3 (LSCF) show much higher electrochemical activity than that of LSM. However, MIEC materials based on cobaltites react readily with YSZ electrolytes to form resistive $La_2Zr_2O_7$ and $SrZrO_3$ phases at the cathode/electrolyte interface. Thus, a thin protective layer such as doped CeO_2 will be needed to inhibit the formation of these resistive phases between the YSZ electrolyte and cobaltite-based cathodes [11]. Recent investigations show that the electrode reaction is more complicated than the mechanism proposed by the TPB model [12]. Perovskite type doped $LaMnO_3$ is the most currently used cathode material for the SOFC, but when it is plasma sprayed very often decomposition occurs [13].

It is known that $La(Sr)CoO_3$ (LSCo) exhibits superior cathodic performance to LSM, e.g., the polarization of an LSCo cathode sputtered on YSZ is very small even at 800°C. On the other hand, LSCo tends to react more readily with YSZ electrolyte than LSM at high temperatures (≥1,000°C), and the resulting $La_2Zr_2O_7$ or $SrZrO_3$ compounds exhibit very high ohmic resistance [14].

Attempts to utilize oxygen-ion conductors such as doped ceria and lanthanum strontium-doped gallates instead of YSZ as a component of composite cathodes have shown some success. However, it is not clear if the high ionic conductivities of these supporting electrolytes are dominant in the comparative enhancement of the cathode activity. One consideration is that the performance of composite cathodes strongly depends on the microstructure; another is that reactions between the supporting electrolytes and the cathode material can negate the benefits of the enhanced oxygen-ion transport [15].

In actual manufacturing of cathode, the processing parameters have significant impact on the microstructural, hence, impact on the SOFC properties. For example, the following processing and microstructural parameters are important in LSM/YSZ cathode [16]:

- The LSM/YSZ ratio of the cathode functional layer (CFL)
- The average grain size and grain size distribution of the CFL and cathode current collector layer (CCCL)
- The thickness of the CFL and CCCL
- The porosity of the CFL and CCCL
- The extent of a graded interface of the CFL
- Production techniques to apply the CFL and CCCL

A major degradation mechanism in SOFC is poisoning of the cathode by chromium from volatilization of the interconnect material. The chromium deposition has been attributed to both chemical and electrochemical mechanisms. For an electrochemical reaction, deposition can occur only where both ions and electrons are available, which, for a purely ionic conducting electrolyte and a purely electronic conducting cathode, can occur only at the three-phase gas–electrolyte–electrode interface. However, the introduction of ionic conductivity into the cathode or electronic conductivity into the electrolyte can allow deposition to occur away from this three-phase interface, and thus alter its effect on the fuel cell performance. In this paper, the chromium poisoning of SOFC cathodes is reviewed, with a focus on the effects of the transport properties of the cathode and electrolyte materials. Manganese can form a $(MnCr)_3O_4$ spinel layer on the scale surface. The reduced chromia activity in the outer spinel phase reduces chromium volatilization and reduces chromium poisoning.

At intermediate temperature, one of the main limiting factors is the high polarization resistance of usual cathode materials, such as LSM oxides, which limits SOFC power densities. Consequently, much attention has recently been focused on improving the cathodic performance by both using mixed ionic and electronic conductor materials and improving the microstructural design at the cathode/electrolyte interface [17].

ANODE

The anode must be porous to allow the fuel to flow to the electrolyte. In an anode supported SOFC, it also provides the mechanical support of the unit cell. Nickel and YSZ (YSZ, $Zr_{0.92}Y_{0.08}O_2$) cermet is the most commonly used material as anode. The purpose of adding YSZ to the anode is to match the CTE between anode and the electrolyte. At the working temperature, which is 700–1,000°C, nickel start sintering and the porosity decreases. This adverse effect will cause the anode to be dense and finally the fuels cannot diffuse through it. The YSZ serves also as the skeleton to prevent the sintering of Ni. The use of nickel is due to its low cost and good electrochemical properties, such as the high electrical conductivity and catalytic activity under reducing conditions. The YSZ also provides an ionic pathway for O^{2-} [18].

The key advantages of this anode are as follows. The conductivity and electrochemical performance with hydrogen fuel is extremely good, consistent with power densities approaching 2 W cm^{-2} at 800°C [19,20]. The processing of NiO–YSZ with YSZ electrolyte layers is straightforward, allowing co-sintering of anode supported structures, and the NiO is easily reduced to Ni providing extra porosity [21]. The reduction of NiO to Ni was achieved in H_2 atmosphere at ~900°C.

There are mainly three categories to prepare an anode support SOFC [43]:

(1) Vapor-phase deposition, such as physical or chemical vapor deposition, spray pyrolysis, etc.
(2) Liquid-phase deposition, such as the sol–gel and liquid-state deposition methods
(3) Particle deposition/consolidation method, such as tape casting, screen printing methods, etc. [22]

Fuel compatibility is an obstacle for SOFC applications. This is a challenging area for anode materials. For example, coal-syngas contains hydrogen sulfide (H_2S) at high concentrations (between 0.5 and 5% H_2S depending on where the coal is mined). The presence of this contaminant in coal-syngas is one of the major obstacles to implementing a coal-syngas SOFC, since an anode catalyst able to handle the presence of H_2S without deterioration over time has not been developed [23]. Barriers to the introduction of hydrogen as the fuel have stimulated interest in developing an anode material that can be used with natural gas under operating temperatures 650°C < T < 1,000°C [24]. A significant portion of the H_2S was transformed to carbon disulfide (CS_2), an important solvent currently used to dissolve heavy hydrocarbon deposits [25] has been observed by using a $La_{0.7}Sr_{0.3}VO_3$ (LSV) anode, and a cathode of $La_{0.85}Sr_{0.15}MnO_3$ (LSM) and YSZ.

Preparation of mixed metal oxide precursors for SOFCs represents a very complex chemical process in which a metal may form oxides, hydroxides, and various complex basic salts as intermediates. Understanding of the relationship between the calcination process, the final composition, particle sizes, sinterability, and SOFC performance for nickel, copper and cobalt-based anode materials is a necessity [26].

In an anode produced by tape casting, powders were dispersed to form homogeneous slurry, which is achieved by milling to get the proper particle size and morphology and to thoroughly mix the components. After tape casting, the anode was dried and presintered at ~1,250°C. The electrolyte layer and the cathode layer are produced by screen printing. Finally, all the layers are sintered together at ~1,400°C. The binders and other organics were burned out during the sintering. After sintering, there could be some distortion in planar SOFC, therefore, a flattening step is usually applied at ~1,360°C with external loading [27].

Liquid mixture technique can be used to obtain the composite precursor. It consists of the evaporation of a dispersion of YSZ powder in a solution of nickel acetate tetrahydrate and ethanol, followed by calcination at 450°C/5 h to eliminate the organic material [28].

It was considered that the morphology and the specific surface area of NiO–SDC composite particles played an important role realizing a high cell performance anode [29]. The effects of precalcined YSZ powders on Ni–YSZ cermet have been investigated. Studies indicated that the crystal structure was not changed, but the porosity of open holes in NiO–YSZ pellets was increased with the precalcining temperature increasing. In addition, the total shrinkage of different anode materials was changed by using different YSZ powders, the value of Ni–YSZ cermet doping with YSZ precalcined at 900 and 1,100°C was closer to that of YSZ electrolyte [30].

Nickel, copper, and cobalt yttria stabilized zirconia (NiYSZ, CuYSZ and CoYSZ) anode materials can be synthesized via hydrolysis of the corresponding chloride solutions with NH_3, NH_3^+ NaOH, and NaOH as precipitation agents. The formation pathway was established for the various products by the direct observation of intermediate species throughout the synthesis process. A comparison of the powders indicates that the choice of precipitation agent greatly influences the final characteristics. The cobalt anodes offered the highest SOFC performance, while within each metal system, the anodes with a crystalline precursor resulted in higher exchange current densities for the charge transfer portion of the impedance spectra [26].

It is also important for fuel cell technology to achieve efficient electrode operation with different hydrocarbon fuels. The Ni–YSZ cermets suffer some disadvantages related to low tolerance to sulfur, carbon build-up when using hydrocarbon fuels (though device modifications and lower temperature operation can avoid this) and volume instability on redox cycling [31]. When using natural gas as fuel for the SOFC, direct internal reforming lowers the requirement for cell cooling and, theoretically, offers advantages with respect to capital cost and efficiency. The high metal content of a nickel/zirconia anode and the high temperature, however, cause the endothermic reforming reaction to take place very fast. The resulting drop of temperature at the inlet produces thermal stresses, which may lower the system efficiency and limit the stack lifetime. To reduce the reforming rate without lowering the electrochemical activity of the cell, the modification, such as wet impregnation procedure for modifying conventional cermets by coverage with a less active metal, could be a promising method to improve the performance of SOFCs with direct internal reforming of hydrocarbons [32].

Direct operation of SOFCs on methane normally degrades nickel anodes because of carbon deposition or coking. Usually, this has been solved by internal reforming, adding large amounts of steam and/or CO_2 into the fuel stream to prevent the carbon formation. Recently, however, it has been shown that SOFCs can also work directly with pure methane without coking, when particular structures of nickel anodes were formulated. Changing the anode composition by adding ceria and operating at higher temperatures were effective in preventing carbon formation. Such results show that nickel anodes can be used directly on methane but the anodes need to be specially formulated in distinct compositions. Another promising way is low temperature reduction of anode [33].

Attention has also been given to Cu-based anode, due to its lower catalytic activity for C–C bond formation, thus suppressing the deposition of carbon on cermet anodes of hydrocarbon-fueled SOFCs [34]. Cu is known to be a poor catalyst for C–C bond breaking and formation. Attempts to prevent Cr oxidation by the electrodeposition of Cu onto the Cr were investigated but the result is not successful in H_2O and H_2 mixture fuels [35].

Other materials such as rare earth-doped ceria and rare earth-doped bismuth oxide have been investigated [36]. Ceria is added because it has both ionic and electronic conductivity under the strongly reducing conditions of the anode, it can be used, either in its pure form or doped with 3+ ions like Gd^{3+}, as the primary anode material. More often, it is added to the anode together with a metal, such as Ni or Cu, either as an interlayer between the electrolyte and the anode 3 or as part of the ceramic metallic– cermet composite itself [37]. Mesoporous flower-like microspheres SOFC with a porous catalyst layer yield much higher maximum power density than those without a catalyst layer. The performance of the fuel cell with a flower-like mesoporous ceria microsphere catalyst layer can achieve a maximum power density up to 0.654 W cm^{-2} at 600°C [38]. The double perovskites that meet the requirements for long-term stability with tolerance to sulfur and show a superior single-cell performance in hydrogen and methane has been reported [24].

Point defects largely govern the electrochemical properties of oxides: at low defect concentrations, conductivity increases with concentration; however, at higher concentrations, defect–defect interactions start to dominate. Thus, in searching for electrochemically active materials for fuel cell anodes, high defect concentration is generally avoided. An oxide anode formed from lanthanum-substituted strontium titanate (La–$SrTiO_3$) in which the oxygen stoichiometry was controlled in order to break down the extended defect intergrowth regions and create phases with considerable disordered oxygen defects. Ti was substituted in these phases with Ga and Mn to induce redox activity and allow more flexible coordination. The material demonstrates impressive fuel cell performance using wet hydrogen at 950°C [31].

Degradation of anode is another major issue for SOFC. Cycling between oxidizing and reducing environments is known to cause anode degradation because of the large Ni–NiO volume change, seal leakage, fuel supply interruption, or system shutdown. Such reduction–oxidation (redox) cycling may occur accidentally and cause catastrophic failure in SOFC systems, especially on the integrity of interfaces. It may also cause significant performance degradation [39]. It would also be desirable for use in small-scale SOFCs if the anodes could endure redox cycling routinely [21].

The degradation has qualitatively been related to microstructural changes in the Ni–YSZ anode of the tested cells [40]. In a symmetrical fuel cell system, the perovskite (LSCM) is shown to be an effective, redox-stable electrode that can be used for both cathode and anode to provide with good performance characteristics [41].

As an alternative to solve the above-mentioned problem of Ni–YSZ anodes, electrically conductive ceramics have been explored, such as $(Ce,Gd)O_2$, Ti–YSZ, Cu_2O, Pr_2O_3, MoO_3, $Sr_{0.6}Ti_{0.2}Nb_{0.8}O_3$, $(La,Ca)TiO_3$, $(La,Sr)TiO_3$, doped $LaCrO_3$ [21].

Another method to suppress the degradation and enhance the long-term and redox stability of the unit cell at elevated temperatures is to use the functional NiO–YSZ composite powder [42]. Ni–YSZ electrode manufactured by reduction of NiO-coated YSZ composite showed good performance and reliability. The anode was manufactured by sintering of the die-pressed NiO–YSZ pellet followed by reduction at 800°C under hydrogen atmosphere. It has a high homogeneity and sufficient contact sites between Ni and YSZ, has an excellent tolerance against thermal and redox cycling.

ELECTROLYTE

YSZ exhibits purely oxygen ionic conduction (with no electronic conduction). The crystalline array of ZrO_2 has two oxide ions to every zirconium ion. But in Y_2O_3 there are only 1.5 oxide ions to every yttrium ion. The result is vacancies in the crystal structure where oxide ions are missing. So, oxide ions from the cathode leap from hole to hole until they reach the anode [1].

Due to a high internal ohmic resistance of the electrolyte, the ionic conductivity decreases with decrease in the operating temperature. The operating temperature of the electrolyte-supported SOFC was limited to 800–1,000°C. Therefore, the electrolyte layer is one of the key components to reach the aims of reducing the operating temperature and increasing cell efficiency. Besides the substitution of the ZrO_2-based electrolyte by higher conductive materials, thickness reduction of the electrolyte is an effective approach to reduce the polarization effects of the layer [43].

Besides YSZ, other materials, such as cerium oxide doped with samarium (SDC), $(Ce_{0.85}Sm_{0.15})O_{1.925}$, cerium oxide doped with gadolinium (GDC), $(Ce_{0.90}Gd_{0.10})O_{1.95}$, lanthanum gallate ceramic that include lanthanum strontium gallium magnesium (LSGM), strontium cerate (SYC), bismuth yttrium oxide (BYO), and barium cerate (BCN) are also investigated.

In the field of SOFCs, an active part of research focuses on finding better electrolytes than YSZ the state-of-the-art material in the field. One of the main purposes is to lower working temperature, which should lessen risks of reactivity with electrode materials. A few years ago, our team has discovered a new family of fast oxide-ion conductors, the so-called LAMOX family, whose parent compound is lanthanum molybdate $La_2Mo_2O_9$. This oxide exhibits, above 600°C, a higher oxide-ion conductivity than YSZ 8%, the best stabilized zirconia, which should result in a decrease of working temperature by more than 150°C. A stabilization of this molybdate against phase transition and reducibility is achieved through appropriate substitution, particularly of molybdenum by tungsten and [44].

INTERCONNECT MATERIALS

The interconnect material's function is to connect the single cells to provide higher output. They are in contact with both the anode and the cathode. Therefore, they must be stable with both electrode materials and in oxidizing and reducing environments. To reduce the ohmic losses and increase efficiency, the interconnect material must also maintain a low electrical resistance. Basically, two materials are used as interconnect materials: metallic and ceramic materials, in which chromium are present, such as $CoCr_2O_4$, $YCrO_3$, and $LaCrO_3$. Glass composite interconnects have also been considered.

Metallic interconnections are better than ceramic because of their low electrical resistivity. Three types of metallic materials are used for interconnection: Fe based, Ni based, and Cr based. Details of them are in [45]. CTE mismatch is a big issue for stainless steel interconnection since the thermal cyclying of SOFC usually generates thermal stresses at the interconnection.

INTERMEDIATE TEMPERATURE (IT-SOFC)

This is a major focus in recent years to make SOFC economically survivable. The operating temperature of IT-SOFC is usually 500–700°C; it can decrease material degradation and prolong stack-lifetime, reducing cost by utilizing metal materials. The disadvantage of IT-SOFC is that the low temperature is usually accompanied by poor activity of the cathode. High-temperature SOFCs are generally all ceramic while IT-SOFCs are metal–ceramic and use stainless steel interconnects instead of more expensive high-chrome alloys or oxides.

Another problem of lowering operating temperature is the increase in the resistivity of electrolyte. There are two approaches to overcome this problem, one is to use alternative electrolyte materials with higher ionic conductivity than YSZ, such as doped ceria and doped $LaGaO_3$, and the other is to reduce the thickness of the YSZ electrolyte [46].

To reduce the resistance of the YSZ electrolyte, electrode-supported in particular anode-supported planar cells with thin film electrolyte of 5–20 μm are often adopted. To operate a SOFC at lower temperatures with high power output, an alternative electrolyte material with high ionic conductivity at low temperatures (500–650°C) should be used. Rare earth oxide-doped ceria (RDC, R is usually Y_2O_3, Gd_2O_3, and Sm_2O_3) is a material of choice due to its superior ionic conductivity, especially in low temperature range of 500–650°C [47].

Besides IT-SOFC, low-temperature (300–600°C) solid oxide fuel cells (LTSOFCs) have also demonstrated promising high performances for different fuels other than only hydrogen. For instance, liquid hydrocarbon fuels, e.g., methanol can be easily thermally decomposed to H_2 and CO that can be directly used for fuel cell operation without an external reformer thus leading to a simple system and highly efficient operation [48].

The Kansai Electric Power Co., Inc. (KEPCO) and Mitsubishi Materials Corporation (MMC) have been jointly developing intermediate-temperature SOFC with the operation temperatures between 600 and 800°C [49]. The electrolyte-supported planar-type cells were fabricated using highly conductive lanthanum gallate-based electrolyte, La(Sr) Ga(Mg,Co)$O_{3-\delta}$, Ni$(CeO_2)_{1-x}(SmO_{1.5})_x$ cermet anode, and Sm(Sr) $CoO_{3-\delta}$ cathode. The 1-kW-class power generation modules were fabricated using a seal-less stack of the cells and metallic separators. The 1-kW- class prototype power generation system with the module was developed with the high performance cell, which showed the thermally self-sustainability. The system included an SOFC module, a dc–ac inverter, a desulfurizer, and a heat recovery unit. It provided stable ac power output of 1 kW with the electrical efficiency of 45% LHV based on ac output by using city gas as a fuel, which was considered to be excellent for such a small power generation system. And the hot water of 90°C was obtained using high temperature off-gas from SOFC.

CONCLUSIONS

SOFC has been demonstrated to be a sustainable alternative energy conversion. Due to its unique structure, materials, mechanism, and working condition, SOFC has many advantages. However, the biggest challenge and barrier for commercialization of the SOFC is the cost. This is because of the manufacturing cost of the ceramic components. Some challenges in the materials of cathode, anode, interconnection, and electrolyte have to be resolved. Material scientists are trying to explore alternative materials to lower the cost. Metal components are easily fabricated but cannot be used at high temperature. Therefore, one direction of the current research on SOFC is to investigate low temperature fuel cells.

References

[1] A.B. Stambouli and E. Traversa, *Renewable & Sustainable Energy Reviews*, 6 (2002) 433–455.
[2] S.C. Singhal, *Solid State Ionics*, 135 (2000) 305–313.
[3] N.P. Brandon, P. Aguiar, and C.S. Adjiman, *Journal of Power Sources*, 147(1–2) (2005) 136–147.
[4] N.M. Sammes, Y. Du, and R. Bove, *Journal of Power Sources*, 145 (2005) 428.
[5] J. Malzbender, T. Wakui, and R.W. Steinbrech, *Fuel Cells*, 6(2) (2006) 123–129.
[6] J. Laurencin, B. Morel, Y. Bultel, and F. Lefebvre-Joud, *Fuel Cells*, 6(1) (2006) 64–70.
[7] H. Yoshida, H. Yakabe, K. Ogasawara, and T. Sakurai, *Journal of Power Sources*, 157(2–3) (2006) 775–781.
[8] T. Suzuki, T. Yamaguchi, Y. Fujishiro, and M. Awano, *Journal of the Electrochemical Society*, 153(5) (2006) A925–A928.
[9] A. Mineshige, et al., *Electrochemical and Solid State Letters*, 9(9) (2006) A427–A429.
[10] K.S. Weil, C.A. Coyle, J.T. Darsell, G.G. Xia, and J.S. Hardy, *Journal of Power Sources*, 152(1) (2005) 97–104.
[11] S.P. Jiang, *Materials Science and Engineering A*, 418(1–2) (2006) 199–210.
[12] A. Barbucci, P. Carpanese, G. Cerisola, and M. Viviani, *Solid State Ionics*, 176(19–22) (2005) 1753–1758.
[13] H. Ageorges, C. Monterrubio-Badillo, T. Chartier, J.F. Coudert, and P. Fauchais, *Surface & Coatings Technology*, 200(12–13) (2006) 3743–3756.
[14] H.Y. Jung, W.-S. Kim, S.-H. Choi, H.-C. Kim, J. Kim, H.-W. Lee, and J.-H. Lee, *Journal of Power Sources*, 155(2) (2006) 145–151.
[15] K. Yamahara, T.Z. Sholklapper, C.P. Jacobson, S.J. Visco and L.C. De Jonghe, *Solid State Ionics*, 176(15–16) (2005), 1359–1364.
[16] V.A.C. Haanappel, J. Mertens, D. Rutenbeck, C. Tropartz, W. Herzhof, D. Sebold, and F. Tietz, *Journal of Power Sources*, 141(2) (2005) 216–226.
[17] M.L. Fontaine, C. Laberty-Robert, F. Ansart, and P. Tailhades, *Journal of Power Sources*, 156(1) (2006), 33–38.
[18] N.Q. Minh, Ceramic fuel cells. *Journal of the American Ceramic Society*, 1993, 76, 563–588.
[19] S. de Souza, S.J. Visco, L.C. De Jonghe, *Journal of Electrochemical Society*, 144 (1997) L35.
[20] A.V. Virkar, J. Chen, C.W. Tanner, and J.-W. Kim, *Solid State Ion*ics, 131 (2000) 189.
[21] S.A. Barnett and B.D. Madsen, *Solid State Ionics*, 176(35–36) (2005) 2545–2553.
[22] P. Von Dollen and S. Barnett, *Journal of the American Ceramic Society*, 88(12) (2005) 3361–3368.
[23] A.I. Marquez, Y. De Abreu, and G.G. Botte, *Electrochemical and Solid State Letters*, 9(3) (2006), A163–A166.
[24] Y.-H. Huang, R.I. Dass, Z.-L. Xing, and J.B. Goodenough, *Science*, 312(5771) (2006) 254–257.

[25] Z. Cheng, S. Zha, L. Aguilar, D. Wang, J. Winnick, and M. Liu, *Electrochemical and Solid State Letters*, 9(1) (2006) A31–A33.
[26] C.M. Grgicak, R.G. Green, and J.B. Giorgi, *Journal of Materials Chemistry*, 16(9) (2006) 885–897.
[27] W. Fischer, J. Malzbender, G. Blass, and R.W. Steinbrech, *Journal of Power Sources* 150 (2005), 73–77.
[28] F.C. Fonseca, D.Z. de Florio, V. Esposito, E. Traversa, E.N.S. Muccillo, and R. Muccillo, *Journal of the Electrochemical Society*, 153(2) (2006) A354–A360.
[29] M. Kawano, et al., *Journal of Power Sources*, 152 (2005) 196–199.
[30] L. Jia, et al., *Journal of Alloys and Compounds*, 414(1–2) (2006) 152–157.
[31] J.C. Ruiz-Morales, J. Canales-Vazquez, C. Savaniu, D. Marrero-Lopez, W. Zhou, J.T.S. Irvine, *Nature*, 439(7076) (2006) 568–571.
[32] R. Dittmeyer and M. Boder, *Journal of Power Sources*, 155(1) (2006) 13–22.
[33] C. Mallon and K. Kendall, *Journal of Power Sources*, 145(2) (2005) 154–160.
[34] E.V. Tsipis, V.V. Kharton, and J.R. Frade, *Journal of the European Ceramic Society*, 25(12) (2005) 2623–2626.
[35] M.D. Gross, J.M. Vohs, and R.J. Gorte, *Journal of the Electrochemical Society*, 153(7) (2006) A1386–A1390.
[36] J.A. Kilner, B.C.H. Steele, Non-stoichiometric Oxides, ed. O.T. Sorensen, Academic Press, New York, 1981, p. 233.
[37] K. Ahn, H. He, J.M. Vohs, and R.J. Gorte, *Electrochemical and Solid State Letters*, 8(8) (2005), A414–A417.
[38] H. Li, et al., *Electrochemistry Communications*, (5) (2006) 833–838.
[39] D. Waldbillig, A. Wood, and D.G. Ivey, *Journal of Power Sources*, 145(2) (2005) 206–215.
[40] T. Klemenso, C.C. Appel, and M. Mogensen, *Electrochemical and Solid State Letters*, 9(9) (2006) A403–A407.
[41] D.M. Bastidas, S. Tao, and J.T.S. Irvine, *Journal of Materials Chemistry*, 16(17) (2006) 1603–1605.
[42] S.-H. Hyun, et al., *Solid State Ionics, Diffusion & Reactions*, 177(9–10) (2006) 931–938.
[43] Y.-Y. Chen and W.-C.J. Wei, *Solid State Ionics*, 177(3–4) (2006) 351–357.
[44] G. Corbel and P. Lacorre, *Journal of Solid State Chemistry*, 179(5) (2006) 1339–1344.
[45] J.W. Fergus, *Materials Science and Engineering A*, 397(1–2) (2005) 271–283.
[46] W. Bao, Q. Chang, and G. Meng, *Journal of Membrane Science*, 259(1–2) (2005) 103–109.
[47] Y.J. Leng and S.H. Chan, *Electrochemical and Solid State Letters*, 9(2) (2006) A56–A59.
[48] B. Feng, C.Y. Wang, and B. Zhu, *Electrochemical and Solid State Letters*, 9(2) (2006) A80–A81.
[49] T. Inagaki, et al., *Journal of Alloys and Compounds*, 408–412 (2006) 512–517.

Chapter 6

BENEFITS AND TEST RESULTS OF A CERAMIC SEPARATOR COMPONENT FOR MICRO FUEL CELLS

Keith Easler
Kyocera America, Inc.

The power demands from portable electronics such as laptop computers, video camera recorders, and personal entertainment push the limits of today's standard battery solutions of Ni-MH and Li-ion based power sources. Recycling of these hazardous materials is also paramount for environmental preservation yet, consumers neglect the impact that these power sources have on our ecosystems. Recycling of these batteries is now becoming common; however, many of these batteries will ultimately reside in landfills upon the exhausting of their usefulness. These demands of environmental preservation and the need for longer life for portable electronics are now driving research for suitable alternatives. The fuel cell is one idea that has the greatest potential to satisfy these growing needs.

The fuel cell has a proven history with solid oxide based systems. Recent needs for miniaturization have generated new ideas of fuel sources and structures. These include direct methanol fuel cells (DMFCs) and polymer electrolyte membrane fuel cells (PEMFCs). These fuel cells have advantages over others for several reasons. A key quality is the potential energy that the fuel sources provide with respect to the fuel costs.

For use in portable electronics, the fuel cell products must be manufactured to fit similar battery form factors. To achieve these mechanical size limitations, fuel cells must be manufactured with materials that permit small features allowing fuel and airflows, electrical connections, and a membrane assembly. In addition, these materials must survive in high temperatures, acidic environments, and remain leak proof. Alumina ceramics have been demonstrated to perform with these requirements using standard multilayer manufacturing techniques employed widely by the semiconductor industry.

This research focuses effort on the development of the structure and features of the ceramic separator and the long-term test results.

Micro fuel cells are primarily designed for portability. Their size is dictated by the size of the device they power. Consumer products such as laptop PCs, portable electronic video games, video cameras, PDAs, and cellular telephones are several examples of portable electronics where a micro fuel cell could be used to power the device. These fuel cells have been designed on primarily two fuel cell types, DMFC and PEMFC. These fuel cell types provide high power density based on their small size and are suitable for powering small devices. A need for high efficiency, combined with a need for lightweight and small size makes these fuel cell types ideal.

These fuel cell designs typically consist of a membrane–electrode assembly (MEA) enclosed with separator plates on either side. The separator plates are designed to function as a fuel supply source to the MEA and collect and distribute the energy generated to the electrical circuits of the fuel cell through anode sides and cathode sides in their respected design. There are two predominant designs using separators plates today: a planar type, where the MEAs are designed in the same plane, and a stacked design, where MEAs are built up on one another. For meeting mechanical packaging limitations, the planar type fuel cells offer advantages for miniaturization. The ceramic multilayer separator plate technology provides an advantage for these fuel cell types because the ceramic can be made thinner than other materials and allow for integrated functionality. The separator plates are designed to incorporate three-dimensional fluid channels for fuel flow and three-dimensional electrical conductors.

A typical PEMFC structure is illustrated in Figure 6-1 as a cross section. The separator plates are manufactured with fluidic channels to supply and distribute fuel to the MEA and collect the electricity generated. Traditionally, these separator plates are commonly manufactured from carbon (graphite) or metal. In those cases, insulation is necessary to prevent corrosion, shorting, and leaking of fuel through the assembly. The use of these materials in the fuel cell, generally require the thickness of the fuel cell to increase. Thus, making it difficult to package in the form factor required for smaller electronic equipment.

Using multilayer ceramic technology, the thickness of the fuel cell is reduced, in part by the use of ceramics fluidic channels and inherent insulative characteristics (Figure 6-2). The fuel channels are incorporated inside the ceramic substrates. This allows the fuel to be protected from contaminates as well as allow for sealing due to ceramics ability to be hermetic when designs require complete sealing. This quality provides a mechanical structure which can effectively supply fuel to the MEA as well as seal off the MEA to optimize efficiency and prevent contamination. In addition, the ceramic separator plates are coated with metals which allow for the interconnection between the cathode and anode sides of the MEA for purposes

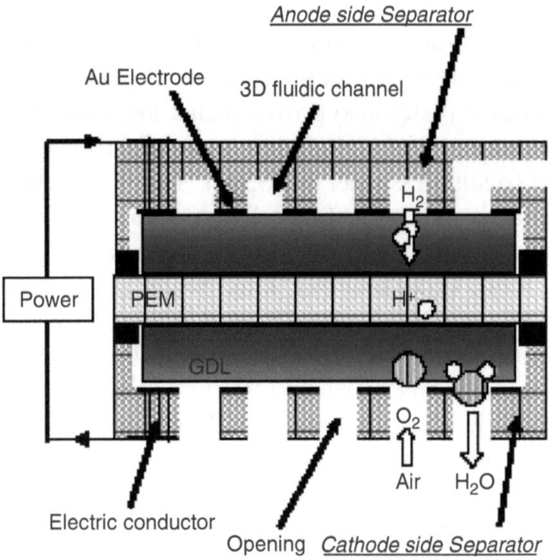

Figure 6-1. Cross-section view of PEMFC with ceramic separators

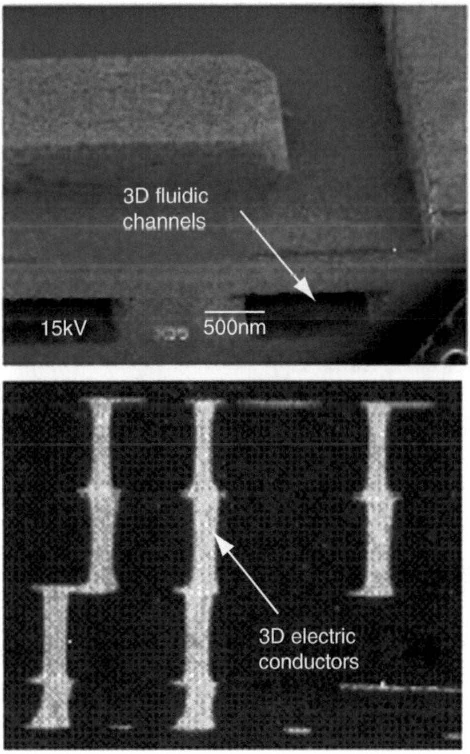

Figure 6-2. Cross-sectional SEM photos of three-dimensional fluidic channels and electric conductors

of collecting the electricity for distribution. These material qualities provide the basis for a thinner structure for the fuel cell, allowing for a reduced size suitable for current form factors for portable electronics.

Ceramic separator plates also provide other key benefits for comparison to alternative separator plates. Ceramic material has a lower bulk density when compared to stainless steel (SUS304). This low bulk density reduces weight and ultimately makes the fuel cell lighter for the consumer. Ceramics are also quite strong. When compared to carbon materials, alumina ceramics are more than two times stronger. This greater flexural strength provides the basis for thinner separator plates when compared to carbon A and carbon B as shown in Figure 6-3. Our comparison illustrates a 65% reduction in thickness when compared to carbon B. Ceramics also have a high young's modulus. This high modulus prevents bending and provides a rigid surface for assembling of fuel cells. This rigidity provides a foundation for gasket and MEA assembly and helps aid in the sealing of these various surfaces. A material property comparison is shown in Table 6-1.

An experiment to determine contact resistance between the gas diffusion layer (GDL) and the separators was designed using the following materials: SUS304, carbon "A," carbon "B," and alumina ceramics. The results are shown in Figure 6-3.

A cross-sectional view of the experiment design is shown in Figure 6-4. This design was used to calculate the resistances of the various separator materials. A separator plate functioning as the cathode side of the fuel cell

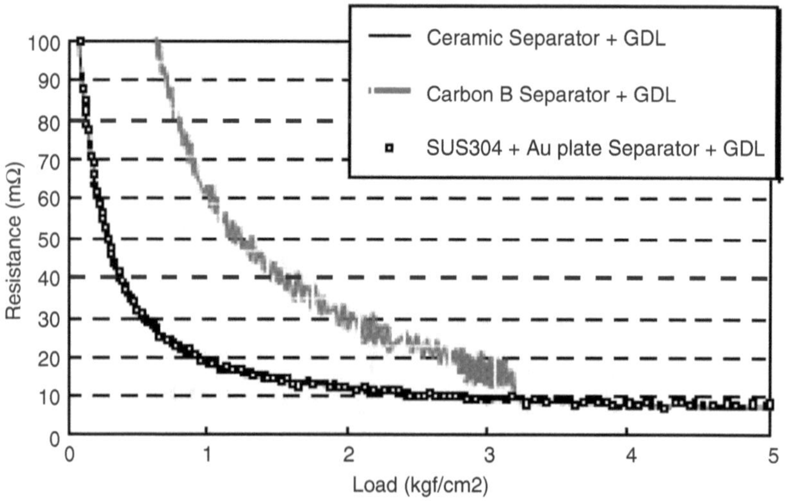

Figure 6-3. Measurement results of contact resistances between the GDL and the separators of each material

Table 6-1. A comparison of material characteristics of separators

Material		Ceramics	Reference			
		Alumina	Carbon A	Carbon B	SUS304	Glass epoxy
Bulk density	(g/cm^3)	3.6	1.95	1.82	7.93	1.8
Mechanical properties	Flexural strength (MPa)	400	43	177	-	-
	Young's modulus (Gpa)	310	21	9	197	20
Process for making fluidic channel		Tape-punching	Molding	Drill	Press, Etching	Laser

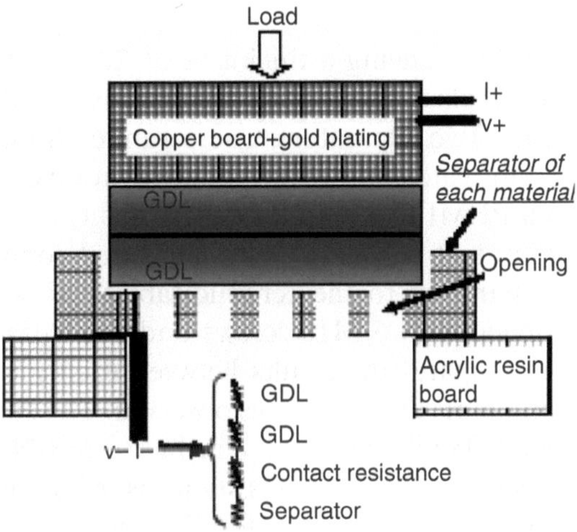

Figure 6-4. Evaluation method to determine contact resistance between the GDL and separator plates under certain loads

was used for each material described above. The goal of this experiment is to understand the contact resistance between the GDL and the separator plates under certain load conditions. A gold-plated copper board was used Total resistance of the assembled unit was calculated by the formula

$$T_R = \mathrm{GDL}_{1R} + \mathrm{GDL}_{2R} + R_C + S_R, \tag{1}$$

where T_R is the total resistance, GDL_{1R} is the GDL 1 resistance, GDL_{2R} is the GDL 2 resistance, R_C is the contact resistance, and S_R is the separator plate resistance.

Figure 6-3 shows the actual resistance measurements under increasing load conditions. The data reveal that contact resistance of the ceramic separator plate is smaller than carbon "B" and equivalent to gold-plated

SUS304. Under load the data show that carbon "B" cracks at a load of 3 kg cm^{-2}, while ceramic and SUS304 remain intact. The rigid characteristics of alumina ceramics permit the separator plate accept high compressive pressures with no deformation and provide gold-plated electrodes to minimize resistance and collect electricity most efficiently.

Further study was performed on the ceramic separator plates to understand the current and resistance characteristics (Figure 6-5). For this experiment, an anodic ceramic separator was developed with meandering fluidic channels to supply dry hydrogen. A cathodic ceramic separator was also developed with rectangular openings for natural air exchange. Using these ceramic separators, three MEAs were connected in series horizontally, using a planar configuration. Low moisturizing MEAs were used with conventional GDLs having a thickness of 200 μm were assembled. For this experiment, a glass epoxy material was also used for a reference comparison material. Two separators, one ceramic and one glass epoxy, were designed and fabricated in the same configurations. These separators were assembled in a PEMFC design for comparison.

The results for contact resistance were measured comparing the glass epoxy fabricated separators to the ceramic fabricated separators. These materials were fastened with 6 M10 screws under increasing torque conditions. Figure 6-6 compares the results between each material under the increasing torque and plots the resistances observed. The contact resistances measured are for one cell under torque and include the GDL resistance. This test shows that the ceramic separator resistance is much lower than that of glass epoxy. The measured resistance for the ceramic was 77% to that of the glass epoxy separator.

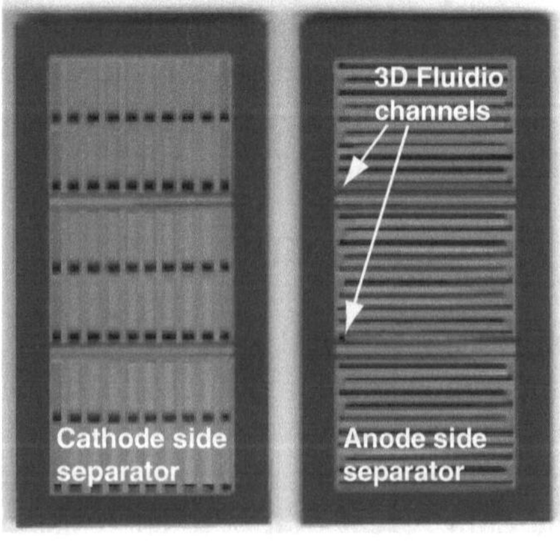

Figure 6-5. Photo of ceramic separators

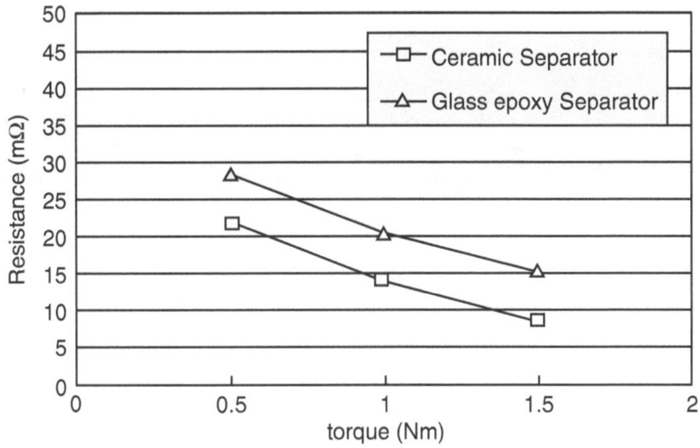

Figure 6-6. Comparison results of contact resistance between a ceramic separator and a glass epoxy separator

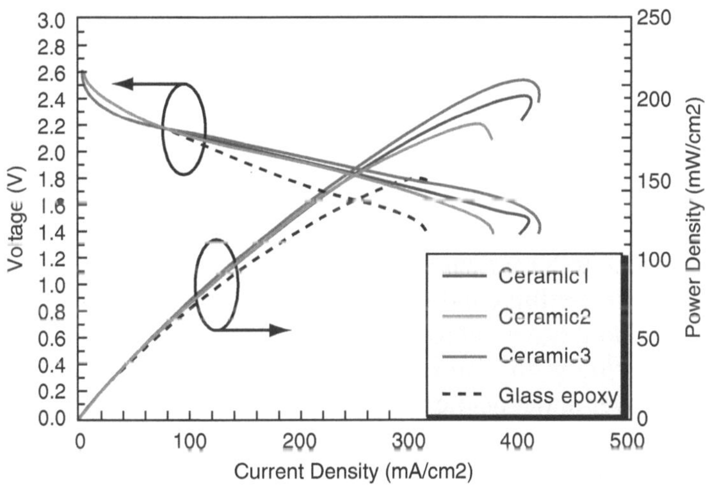

Figure 6-7. Results of current and voltage measurements

Further comparative study reviewed the maximum power density for ceramic separators and glass epoxy materials (Figure 6-7). The results indicated that the planar PEMFC using ceramic separators achieved over 200 mW cm^{-1} at maximum power density. Compared to glass epoxy, the ceramic separator maximum power density was 138% higher.

Long-term stability tests were also conducted on the ceramic separators. These tests evaluated the corrosion resistance during continuous power generation of a single cell PEMFC. The separators used in the

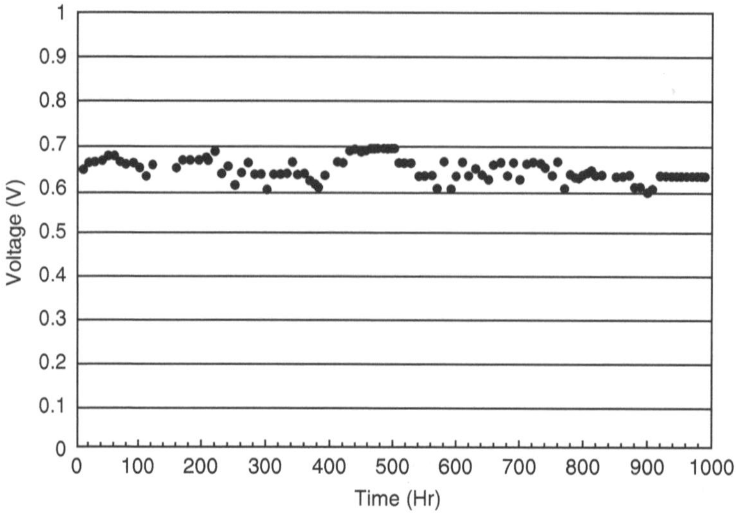

Figure 6-8. Long-term stability results of ceramic separators

experiment were designed with meandering flow channels for dry hydrogen and dry air. Figure 6-8 shows the long-term stability test voltage output results of the PEMFC using the ceramic separators.

These test cells were operated and monitored for 1,000 h. The test cells were maintained to have a current density of 200 mW cm^{-1} throughout the test period with testing conducted at room temperature. Voltage output was evaluated and observed to within ±10% over the duration of 1,000 h. Long term testing results proved that ceramic separators were capable of generating power under acidic conditions. This test will be performed until 3,000 h are reached.

The results for using ceramic materials as an integral component for micro fuel cells have conclusive evidence that ceramics make an excellent separator plate for micro fuel cells. The thin structure of ceramics allows for rigidity without sacrificing increased resistance. A lower contact resistance when compared to other common separator materials increases power output. Ceramic's inert composition also provides excellent reliability in acidic conditions of the fuel cell. Ceramic separator plates achieved the highest power density of comparable materials and subsequently produced less than 10%± voltage variation over 1,000 h of testing.

The ceramic separator provides groundwork for future power sources that enable small form factors for portable electronics.

ACKNOWLEDGMENTS

Special thanks are given to several individuals at Kyocera Corporation Research and Development in Kagoshima, Japan, including Mr K. Onitsuka, Mr S. Satou, Mr H. Otomaru, and Mr T. Miyao. Also a special acknowledgment is given to the members of Fraunhofer Institute of Solar Energies Systems for their advice for evaluations and product development.

Chapter 7

FOIL TYPE MICRO PEM FUEL CELL WITH SELF-BREATHING CATHODE SIDE

Stefan Wagner, Robert Hahn, and Herbert Reichl
Fraunhofer Institute Reliability and Microintegration (Fraunhofer-IZM), Gustav-Meyer-Allee 25, 13355 Berlin, Germany

INTRODUCTION

Despite increasingly efficient components and low power technologies, the energy demands of portable electronic products such as next-generation mobile phones, wearable computers, autonomous sensors and microsystems will rise dramatically in the future due to their growing functionality. As improvements in battery technology have so far been limited to energy density increases of only a few percent per annum, over the past few years many R&D activities have concentrated on alternative forms of portable power supply. One of the most promising candidates is micro fuel cells (FCs) based on polymer electrolyte membranes (PEMs). They could be used as energy sources for applications in devices such as hearing aids, chip cards, wireless sensor networks and other small portable devices, thereby replacing Li-polymer batteries, button cells and zinc air batteries. Compared to batteries the environmental impact of fuel cells is much lower [1].

FUEL CELL PRINCIPLE

Since a single fuel cell has an operation voltage of ca. 0.5 V, a multitude of cells are needed which are typically assembled in stack configuration.

A fuel cell stack can be subdivided into three constituent component groups: the membrane–electrode assemblies (MEAs) which fulfils the electrochemical function of the fuel cell, the bipolar plates (commonly consisting of graphite and carbon-filled polymers), which supply the MEAs with hydrogen and oxygen, providing cooling and conductive electronic paths, and the gas diffusion layers (GDLs), which are inserted between MEAs and bipolar plates to distribute the reactants uniformly. The GDL consists of carbon fabric and Nafion®. Current is produced across the surface of the MEAs and collected at the ends of the stack. Fuel cells of the air-breathing type are using ambient air as an oxidant and hydrogen as fuel.

PEM fuel cells operate with hydrogen. At the moment, there is no hydrogen storage available which is suitable for miniature applications. For direct methanol fuel cells (DMFCs) a better storage opportunity exists in form of methanol cartridges.

WHY MEMS FUEL CELLS?

There are mainly two driving forces behind this development:

1. Using new technologies and designs it should be possible to significantly improve fuel cell performance when micro-scale phenomena are exploited. However, such benefits can only be realised if the fuel cell devices can be fabricated using available manufacturing techniques. They are in most cases adapted from semiconductor and microsystems technology. Fuel cells built to exploit micro-scale phenomena would be smaller, make better use of volume and could obtain improved heat and mass transfer.
2. The majority of research on micro-scale fuel cells is also aimed at micro-power applications. There are many new miniaturised applications which can only be realised if a higher energy density power source is available compared to button cells and other small batteries. Miniaturisation of the conventional fuel cell stack technology is not possible down to these dimensions [2].

MEMS APPROACHES TO MICRO FUEL CELLS

Most of the approaches for MEMS fuel cells comprise basic research and materials development. Examples are the development of microporous substrates, the polymerisation of the ionomer directly on the cell's structure and subsequent coating with catalyst. These approaches have in common that a great effort is needed to establish the process and to optimise the electrokinetic performance of the catalyst layer. Adapting these coating processes to a new cell geometry might lead to high set-up cost as new coating tools have to be manufactured.

In the domain of MEMS-based miniature FC, various solutions have been reported. Since the performances do not seem to be the better criteria to classify FC – actually we can notice in the literature that the power densities of the miniature FC range from μW cm^{-2} up to mW cm^{-2}. The structures are often the same, two micro-machined plates for fuel delivery and current collectors sandwiching a membrane–electrodes assembly (MEA). One can index the different MEMS-FC by similar materials. In the following, it will not be focused on the different fuels used (hydrogen for proton-exchange-membrane FC or PEMFC, methanol for direct methanol FC or DMFC, ethanol, formic acid) but it will be precised for each FC described the fuel supply conditions.

SILICON-BASED FC

As the base material for MEMS technologies, Si is also the most common material encountered in MEMS-based FC. Its properties and the micro-fabrication techniques associated to it such as photolithography, wet or dry etching, depositing (sputtering, CVD, thermal oxidation, etc.) are now well known and mastered. Another advantage of Si-based FC may also be to facilitate the possible integration of the FC with other electronic devices on the same chip.

The Tohoky University, Japan (Min et al.) reported a μ-PEFC using micro-fabrication techniques [3]. This FC is based on two different designs, the "alternatively inverted structure" and the "coplanar structure". These structures use Si substrates with porous SiO_2 layers with Pt-based catalytic electrodes and gas feed holes, glass substrates with micro-gas channels and a polymer membrane (Flemion® S). In spite of a reported enhancement [4], the FC reached poor results, the power density is only 0.8 mW cm^{-2}.

The Stanford University (Lee et al.) proposed a 'flip-flop' μ-FC design where both electrodes are present on the same face [5]. If this design does provide ease of manufacturing by allowing in-plane electrical connectivity, it complicates the gas management. Instead of electrons being routed from front to back, gasses must be routed in crossing patterns, significantly complicating the fabrication process and sealing. Peak power in a four-cell assembly achieved was still 40 mW cm^{-2} with H_2 as fuel.

Meyers et al. from Lucent Technologies have proposed two alternative designs using Si – a classical bipolar using separate Si wafers for the cathode and the anode and a less effective monolithic design that integrated the two electrodes onto the same Si surface [6,7]. In the bipolar design, both electrodes were constructed from conductive Si wafers. The reactants were distributed through a series of tunnels created by first forming a PS layer and then electropolishing away the Si beneath the porous film. The FC was completed by adding a catalyst film on top of the tunnels and finally by casting a Nafion® solution. Two of these membrane–electrode structures were made and then sandwiched together. A power density of 60 mW cm^{-2} was announced for the bipolar design with H_2 supply.

Yu et al. from the Hong Kong University described a miniature FC consisting in a MEA between two micro-machined Si substrates [8]. By sandwiching Cu between layers of gold, they were able to decrease the internal resistance of the thin-film current collectors, which involved an increase performance of the FC, achieving a peak power density of 193 mW cm^{-2} with H_2 and O_2.

Yen et al. [9] (University of California/Pennsylvania State University) presented a bipolar Si-based micro-DMFC with an integrated MEA (Nafion®-112 membrane) and 750 μm wide and 400 μm deep micro-channels fabricated using Si micro-machining (DRIE). This μDMFC with an active area of 1.625 cm^2 has been characterised at near room temperature (RT), showing a maximum power density of 47.2 mW cm^{-2} at 60°C when 1 M methanol was fed but only 14.3 mW cm^{-2} at RT. Since then, Lu et al. [10] have enhanced the performance of the μDMFC to a maximum power density of 16 mW cm^{-2} at RT and 50 mW cm^{-2} at 60°C with 2 M and 4 M methanol supply with a modified anode backing structure enabling to reduce methanol crossover.

The University of South Florida (Aravamudhan et al.) presented an FC powered by ethanol at RT [11]. The electrodes have been fabricated using macro-porous Si technology. The pores developed act both as micro-capillaries/wicking structures and as built-in fuel reservoir, reducing the size of the FC. The pore sizes dictate the pumping/priming pressure in the FC. The PS electrode thus eliminates the need for an active external fuel pump. The structure of the MEA consists of two PS electrodes sandwiching a Nafion®-115 membrane. Pt was deposited on both the electrodes micro-columns to act both as an electrocatalyst and as a current collector. The FC reached a maximum power density of 8.1 mW cm^{-2} by supplying 8.5 M ethanol solution at RT.

The University of Illinios (Yeom et al.) demonstrated the fabrication of a monolithic Si-based micro-scale MEA consisting of two Si electrodes, with catalyst deposited directly on them, supporting a Nafion®-112 membrane in-between [12]. The electrodes are identically gold-covered for current collecting, and are covered with electrodeposited Pt black. The electrodes and the Nafion® membrane are sandwiched and hot-pressed to form the MEA. The complete fuel cells have been tested with various fuels: H_2, methanol and formic acid and reached 35 mW cm^{-2}, 0.38 mW cm^{-2} and 17 mW cm^{-2}, respectively, at RT with forced O_2. More recently, performances with formic acid as fuel were increased up to 28 mW cm^{-2} with electrodeposited Pd-containing catalyst at the anode [13].

The Carnegie Mellon University (Yao et al.) is currently working on a RT DMFC to produce a net output of 10 mW for continuous power generation [14]. Their works focus on the design of the complete system including water management at the cathode, micro-pumps and valves, CO_2 gas separation and other fluidic devices. A passive gas bubble separator removes CO_2 from the methanol chamber at the anode side. The back planes of both electrodes are made of Si wafers with an array of etched micro-sized holes. Nano-tube catalysts are fabricated on the planes. A 3% methanol solution at the anode and the air at the cathode are driven by natural convection instead of being pumped. A micro-pump sends water back to the anode side. With 25 mW cm^{-2}, the total MEA area around 1 cm^2 can provide enough power to a 10 mW micro-sensor along with the extra power needed for internal use, such as water pumping, electronic controls and process conditioning.

The French Nuclear Research Center CEA announced the successful fabrication of high performances prototypes fuelled by H_2 and based on thin films type structures on Si substrate obtained by micro-electronics fabrication techniques (RIE for fuel micro-channels, PVD for anode collector, CVD, serigraphy, inkjet for Pt catalyst, lithography) with a Nafion® membrane [15]. An impressive power density of 300 mW cm^{-2} with a stabilisation around 150 mW cm^{-2} during hundred hours was reported.

The FEMTO ST Institute [Pichonat et al.] demonstrated an alternative solution which does not use an ionomer for the proton exchange membrane but consists in a porous silicon membrane with a proton conducting silane grafted on the pores walls [16,17]. A demonstrated fuel cell with dimensions of 8×8 mm^2 and an active area of 7 mm^2 with this membrane achieved a maximum power density of 58 mW cm^{-2} at room temperature with H_2 as fuel.

METAL-BASED FC

Less expensive than silicon and easier to assemble with other FC components, micro-machined metal foils are also one possible choice to obtain low-cost functional FC.

Using the same structure as previously described [9,10], Lu et al. have recently replaced the Si judged too fragile for compressing and good sealing with the MEA by very thin stainless steel plates as bipolar plates with the flow field machined by photochemical etching technology [18]. A gold layer was deposited on the stainless steel plates to prevent corrosion. This enhanced FC reached 34 mW cm^{-2} at RT and 100 mW cm^{-2} at 60°C.

The IMTEK (Müller et al.) used micro-machined metal foils to form the flow fields of their μ-FC [19]. Using Gore MEA, they were able to form very thin, high power density stacks. They demonstrated both uncompressed and compressed FC designs. The uncompressed one had a peak power density of 20 mW cm^{-2} and the 5-cell compressed stack achieved 250 mW cm^{-2}.

Another work of the University (Ha et al.) of Illinois deals with metallic miniature FC structure [20]. They described the design and performance of a passive air breathing direct formic acid FC (DFAFC). The MEA was fabricated in house with catalyst inks directly painted on a Nafion®-117 membrane. The current collectors were fabricated from Ti foils electrochemically coated with gold. The miniature cell at 8.8 M formic acid produced a maximum power density of 33 mW cm^{-2} with pieces of gold mesh inserted between the current collector and the MEA on both sides of the MEA. With Pd black used as the catalyst at the anode side [21], they recently obtained a maximum power density of 177 mW cm^{-2} at 0.53 V for their passive DFAFC with 10 M formic acid.

POLYMER-BASED FC

A third way to build up μFCs based on MEMS-polymers such as polydimethylsiloxane (PDMS) or polymethyl methacrylate (PMMA) or PCB-materials such as polyimid (PI) or FR4. These polymers can be micro-machined by molding or by laser ablation. Shah et al. [22,23] have developed a complete PEMFC system consisting of a PDMS substrate with micro-flow channels upon which the MEA was vertically stacked. PDMS micro-reactors were fabricated by employing micro-molding with a dry etched silicon master. The PDMS spin coated on micro-machined Si was then cured and peeled off from the master. The MEA employed consisted in a Nafion®-112 membrane where they have sputtered Pt through a Mylar mask. Despite an interesting method, this FC gave poor results, a power density of 0.8 mW cm^{-2} was achieved.

The Nanyang University (Chan et al.) works on the fabrication of a polymeric μPEMFC developed on the basis of micro-machining of PMMA by laser [24]. The micro-channels for fuel flow and oxidant were ablated with a CO_2 laser. The energy of the laser beam has a Gaussian distribution thus the cross section of the channel also has a Gaussian shape. A 40 nm gold layer was then sputtered over the substrate surface to act both as the current collector and corrosion protection layer. The Gaussian shape allows gold to cover all sides of the channel. In this μ-FC, water generated by the reaction was utilised for gas humidifying. The flow channel has a spiral shape which enables the dry gas in the outer spiral line to become hydrated by acquiring some of the moisture from the adjacent inner spiral line. The MEA consists in a Nafion®-1135 membrane with a hydrophobic carbon paper and a diffusion layer (carbon powder and PTFE) on one side and coated with a catalyst layer (ink with Nafion®-112 solution and Pt-carbon) on the other side. Silver conductive paint was printed on the other side of the carbon paper to increase its conductivity and to contribute to collect current. In the final step, the two PMMA substrates and the MEA were bonded together using an adhesive gasket. H_2 was supplied by hydride storage, the air by an air pump, all tests were performed at RT. A high power density of 315 mW cm^{-2} at 0.35 V has been achieved.

Schmitz et al. [25] demonstrated a planar fuel cell design based on printed circuit board (PCB) technology. This planar design consists of an open cathode side which allows a completely passive, self-breathing operation. Power densities of 100 mW cm^{-2} at 500 mV with hydrogen were achieved. Long-term operation for more than 1,500 h has been demonstrated.

FOIL TYPE MICRO FUEL CELL DESIGN

Planar Design of PEMFC

For a successful integration the dimension of the fuel cell must be in accordance with the existing geometries of the electronic device. A flat design of fuel cells fulfills these needs in most cases. As the fuel cell in this packaging concept serves as part of the housing, the volume needed for the fuel cell as power source is minimised.

A planar design allows a passive, full self-breathing operation of the fuel cell. Additional fans or air movers are not needed.

In order to achieve higher voltages, a plurality of cells has to be serial connected whereas the cathode is connected with the anode of the following cell.

In our design the interconnection between cells is performed outside the membrane area which reduces sealing problems.

Micro Flow Field

For this concept it is necessary that the fuel cells are constructed without additional GDLs because by using a continuous piece of GDLs a short circuit will occur between adjacent cells. In order to avoid this, a separate GDL for each single cell must be assembled, which is contrary to the idea of processing thousands of fuel cells at the same time using wafer and foil processing. Therefore in this design concept diffusion layers are not used.

The function of a conventional diffusion layer assumes a micro-structure which is merged of the anode and cathode side. The micro-structures used in the described setup are in the dimension of the carbon fibers in conventional diffusion layers. Moreover, the micro-structures serve as flowfield and are designed to assure a homogenous gas supply all over the MEA.

The planar design consists of an open cathode side for oxygen supply from the air and water removal. A tight contact of the ribs of the flowfield with the MEA is achieved by using a bending force according to Figure 7-2 or by adhesive bonding. Several different designs for the micro-flowfields were analysed. The anode micro-flowfield was patterned with dimensions down to 10 µm.

Due to the poor electrical lateral conductivity of the catalyst layer on the MEA, a voltage drop occurs in the MEA in a direction perpendicular to the gas channel. Therefore the channel needs to be quite narrow in order to minimise resistive losses. As the ribs cover the catalyst layer, fuel trans-

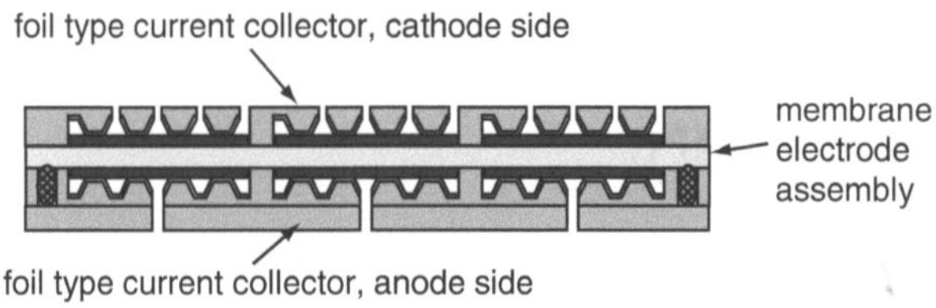

Figure 7-1. Cross-sectional view of the Fraunhofer IZM micro fuel cell

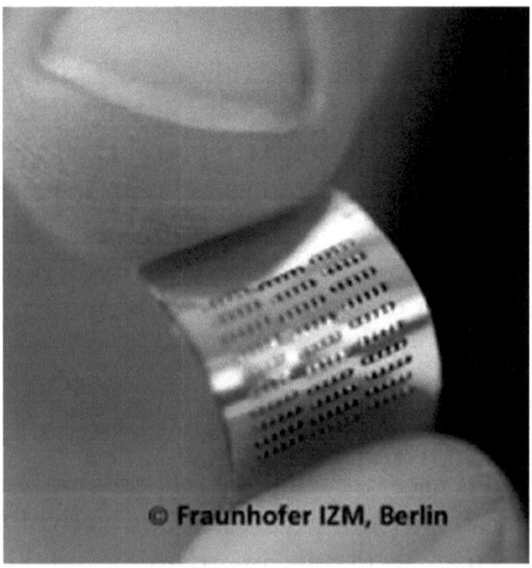

Figure 7-2. Foil type micro fuel cell demonstrator

port under the ribs is limited. These effects were evaluated with measurements of varying channel geometries and are discussed with the aid of an analytical model [25]. An initial optimisation of the aspect ratio between the channels and ribs of the flowfield was carried out. A conducting layer is deposited on the flowfield which serves as current collector.

The cathode side is realised as a micro-structured and perforated current collector foil which allows a self-breathing operation.

The entire fuel cell has a thickness of less than 200 μm. Figure 7-1 shows a schematic cross-sectional view of the foil type micro fuel cell concept and in Figure 7-2, the demonstrator.

TECHNOLOGY

Two different developments are tracked by the Fraunhofer IZM on the field of micro fuel cells.

On one side the development is based on thin film and micro-patterning technologies. Wafer level and foil processes used to produce high density interconnect electronic modules, and wafer level packaging was adapted to micro fuel cell development to achieve the required miniaturisation and cost reduction. By using reactive ion etching, high aspect ratio capillary structures of the anode and cathode side flow fields were achieved.

The foil-type planar micro fuel cell technology is based on the following key technologies:

- Sandwich laminate of polymer–stainless steel foils
- Lithography and patterning of free standing grid micro-structures
- Micro-patterning of flow fields
- Subtractive patterning of MEA electrodes
- Adhesive sealing and electrical interconnection

And on the other side the development is based on high density PCB technologies.

Foil Technology

The micro fuel cell itself consists of polymer and metal foils enabling the opportunity for mass production processes such as reel to reel processes.

The current collector foils with integrated flow field of the anode and cathode side consist of metalised and micro-structured sandwiched metal–polymer foils.

The flow field, sealing joint and gas inlets are fabricated at the anode side, the openings for the air inlet are fabricated at the cathode side by micro-patterning. The metal foils act as the housing of the fuel cell. Polymer layers are laminated on the metal foils which are fixed on a Si wafer. This wafer is only used as a carrier substrate. The lamination processes are developed for a film thickness less than 100 µm.

The metal substrate has a thickness between 10 and 30 µm with openings for the gas and air inlets. These openings are structured in a prior process by means of wet etching or laser processing.

DuPont Pyralux® coverlay act as polymer foil. The composites are made of DuPont Kapton® polyimide film, coated on one side with a proprietary B-staged modified acrylic adhesive.

The flow field and sealing joint is patterned into the polymer layer by reactive ion etching (RIE). Investigations with different process gases and process parameters were accomplished. In addition to the flow field structure the hard mask contains a structure for the sealing joint.

After fabrication of the micro-channels in the flow field, another lithographic process step is necessary to metallise the current collector. Electroplating is used to reach the necessary thickness. Figure 7-3 shows the process sequence for patterning of the anodic flow field. Different flow field designs with regard to structure depth and width were realised. Figure 7-4 shows SEM pictures of a flow field pattern. Figure 7-5 points out a rip of the flow field structure in detail. The openings at the cathode side current collector, which are essential for the self-breathing function, are structured by RIE as well. A perforated metal foil is used as dry etch mask and acts as outer surface afterwards. At the back side of the foil a grid of current conductors is fabricated using the same semi-additive process as for the anode current collector. In both cases electroplated Au is used. Different dimensions of these openings are realised. Figure 7-6 shows the cathode collector foil and Figure 7-7 gives an overview of RIE etch rates.

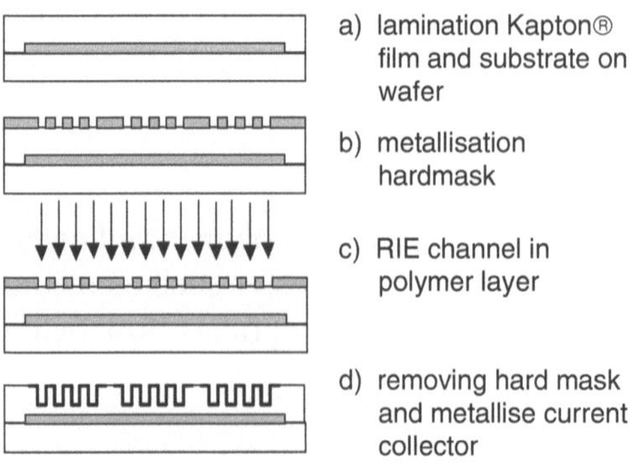

Figure 7-3. Schematic process sequence for the patterning of the anode

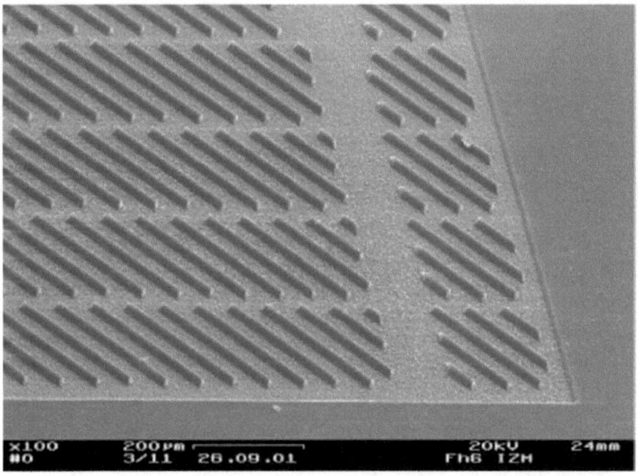

Figure 7-4. SEM pictures of anodic side flowfield

Figure 7-5. SEM pictures of anodic side flowfield detail

PCB Technology

The main motivation for the realisation of micro fuel cells in printed circuit board technology is to achieve low costs by using mass production processes. These technologies make a reel-to-reel processing possible.

The base substrate is FR4 or PI with a copper layer with a thickness of 70 µm. The polymer layer has thickness of 50–100 µm. In the Cu layer the flow field structure is patterned. This patterning technology is a two-stage etch process. In the first step the channels for the H_2 distribution were realised. In the second step the current collector structure was patterned.

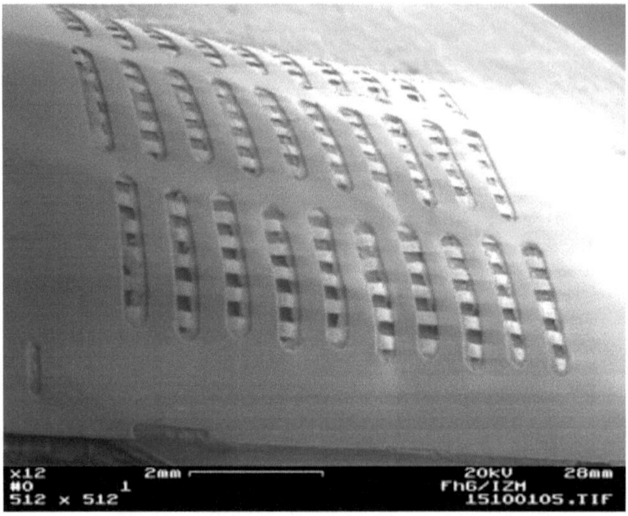

Figure 7-6. SEM pictures of cathodic side current collector

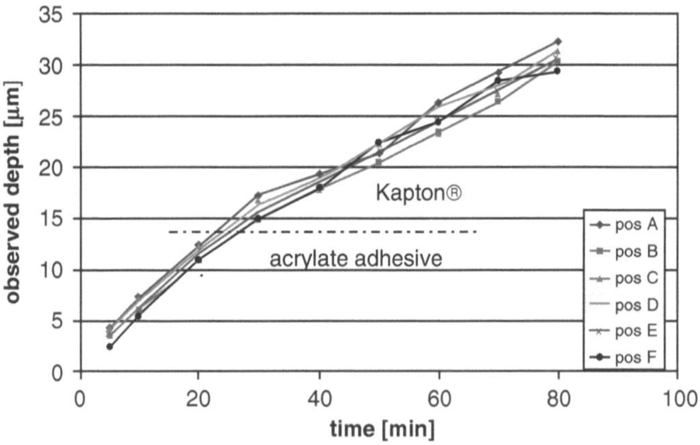

Figure 7-7. Etch rate in Kapton® foil

In both cases a thin tin layer serves an etch mask. Afterwards an NiAu layer was deposited on the Cu to prevent corrosion. In a last process step the openings for the H_2 in- and outlet on the anodic substrate and the openings for the self-breathing function on the cathodic substrate are realised by a LASER process.

These substrates for the anode and cathode side have been assembled with the MEA.

MEA Segmentation

By using a continuous piece of MEA for several cells in a planar arrangement, internal bypasses due to the lateral electrical conductance of the catalyst layer may occur [26]. This effect must be avoided by structuring electrically insulated segments in the catalyst layer.

By the use of conventional coating processes for MEAs a segmentation of catalyst layer can be manufactured. Nevertheless with these coating technologies the distances between the electrode areas are in the range of some millimetres. Realising a miniaturised planar fuel cell with a plurality of serial connections means that a relative large inactive area is wasted between each cell. If several cells have to be serial connected on small area in the range of 1 cm^2 the conventional segmented coating is not appropriate. Figure 7-8 shows a schematic picture of the fuel cell with serial interconnection of three cells in planar configuration.

For these investigations segmentation was realised by reactive ion etching (RIE) and LASER ablation.

With these technologies the distances between the segments can be reduced to less than 200 μm.

The micro- and nano-sized particles of graphite and platinum found in the catalyst layer reach absorption values in the UV and NIR of up to 99%. Due to the high absorption the catalyst layer is removed very efficiently and selectively by LASER ablation. The ionomer material of the membrane is partially transparent for the LASER radiation. Thus the catalyst layer can be removed on both sides of the MEA at the same time. The figure 7-9 shows the segmentation gap in the catalyst layer of a MEA.

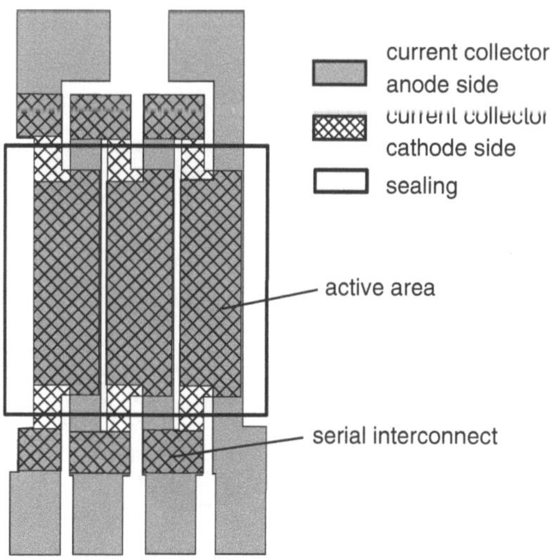

Figure 7-8. Serial interconnecting configuration

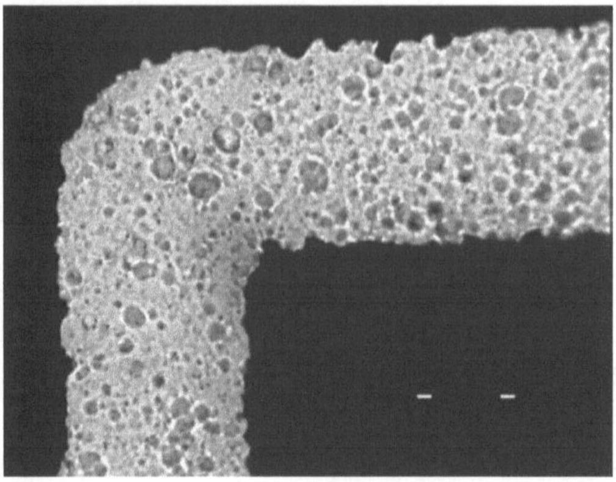

Figure 7-9. The ablated catalyst layer on the MEA. Patterning with RIE was also studied extensively but resulted in high processing times since overheating of the MEA has to be avoided

Assembly

Contacting pads from the cells to be interconnected overlap in the area outside the membrane. Thus a serial connection is realised by merging the overlapping pads by the use of gang bonding technologies or screen printing electrical conducting adhesives. A dispensing technique or screen printing was used for integrated planar sealing. The principle is demonstrated in Figures 7-10 and 7-11. A multitude of fuel cells can be assembled at the same time.

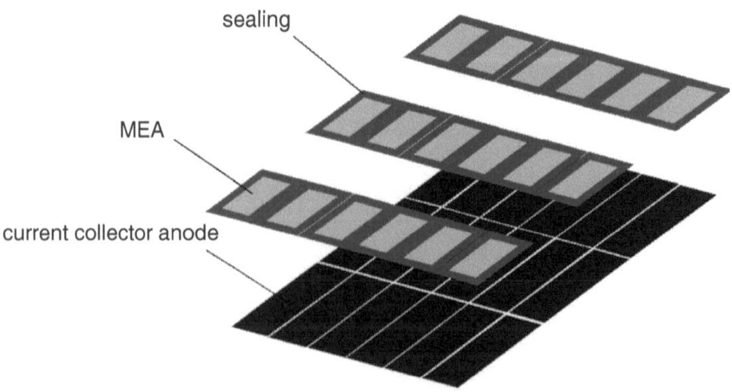

Figure 7-10. Assembling of micro fuel cells – MEA to anodic current collector

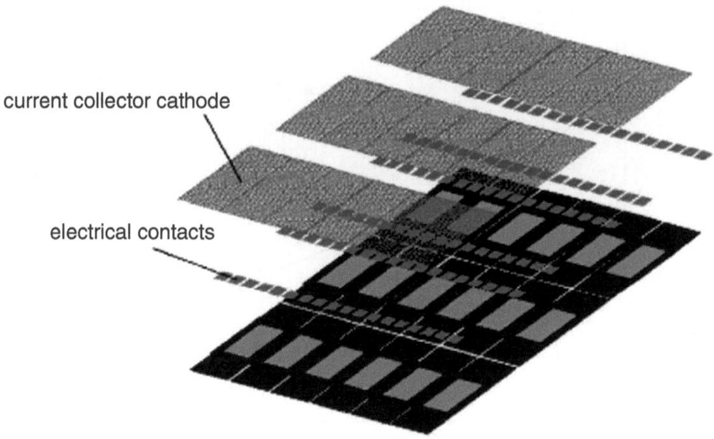

Figure 7-11. Assembling of micro fuel cells – cathodic current collector

ELECTRICAL CHARACTERISATION

The cell prototypes are electrically characterised by *V/I* measurements. The measurements were carried out at room temperature and under natural air convection of the cathode. All measurements were made with dry hydrogen 3 sccm hydrogen flow and no additional humidification.

Other measurements were conducted in an environmental chamber at a variety of temperatures and relative humidity. In the climate chamber the fuel cells were placed in an extra enclosure to keep off any air movement produced by the circulation system. Figure 7-12 shows the *V/I* characteristic of a single cell with an active area of 0.54 cm^2.

Figure 7-13 shows the *V/I* characteristic of different designs of the anodic flow fields. Both cells, which are referred to as designs 1 (meander anodic flow field) and design 2 (parallel rib flow field) have the same cathode current collector with 400 μm openings. The anodic flow field of design 1 has a meander shape with 250 μm wide channels while design 2 has a rib structure and 100 μm wide channels. The structure depth is 20 μm. As can be seen from Figure 7-5 there is only a minor difference in performance between both types.

In another experiment the depth of the anode flow field was varied between 25 and 40 μm. Here the differences of the *V/I* curve is within the measurement error. Cells with lower channel depth show higher power fluctuations during long-term tests.

Figure 7-14 gives an example of long-term tests in the climate chamber. At high temperatures (60°C) the cell performs best at high humidity while

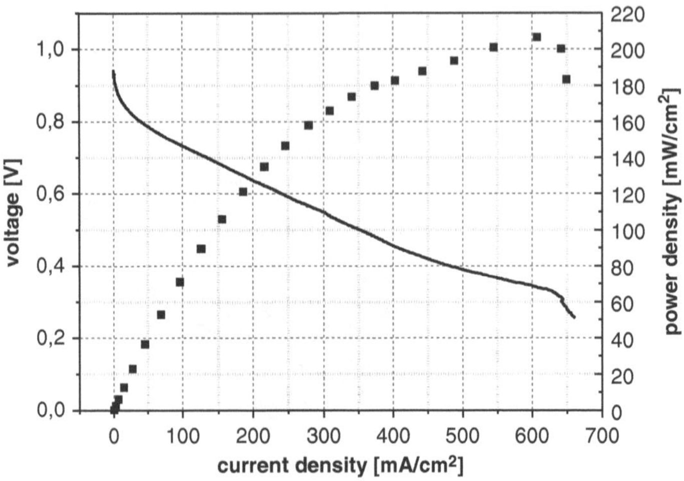

Figure 7-12. VII characteristic of a single micro fuel cell

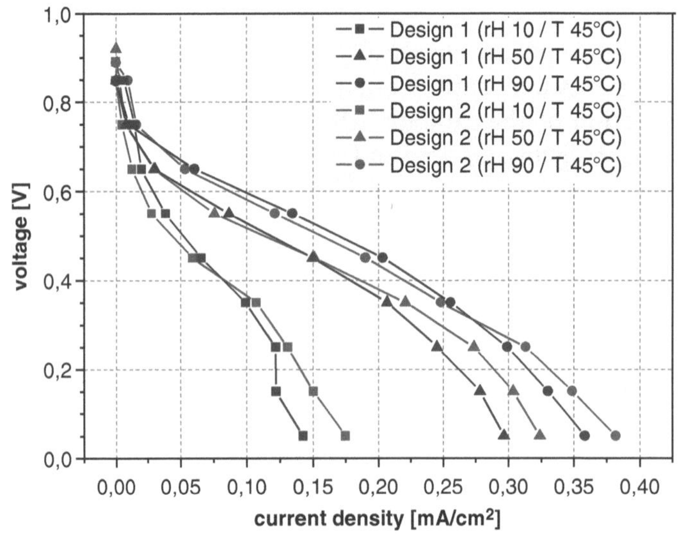

Figure 7-13. Characteristic of PEM micro fuel cells. Design 1: meander anodic flow filed, design 2: parallel rib flow field

the optimum at low temperatures (10°C) lies at lower humidity. Compared to conventional fuel cells with GDLs the micro fuel cell responds at slightly lower temperatures to dry ambient conditions (membrane dry out) compared to cells with GDL. On the other side, at high humidity and low temperatures the micro fuel cell current density drops slightly faster.

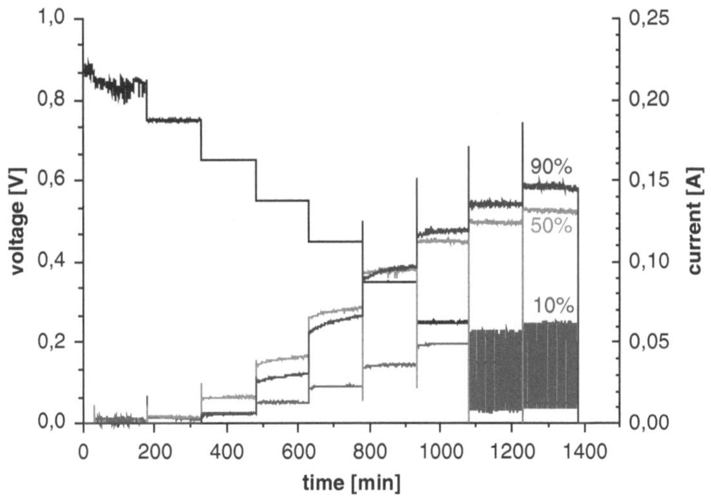

Figure 7-14. Characterisation of micro fuel cell at varying ambient humidity and 60°C at potential steps

At low and medium relative humidity of the air, the performance of the meander design is slightly superior to the rib structure as can be seen in Figure 7-13.

In summary, stable operation of the micro-cells was achieved at 80 mW cm^{-2} in long term tests at varying air conditions. A power density of 120 W cm^{-2} can be achieved in a narrow climatic range. The fuel cells perform best at normal ambient conditions.

MEA Segmentation and Serial Interconnection

As outlined above, the segmentation of MEA electrodes is a key process to implement the concept of integrated fabrication of micro fuel cells which consists of merely three foils. Figure 7-15 shows the electrical characterisation of the serial interconnected three cell demonstrator. Approximately 40 mA can be drained at 1.5 V. To illustrate the importance of MEA patterning a fuel cell has been assembled with three serial interconnected current collector structures according to Figure 7-9 but with a continuous piece of MEA.

Figure 7-16 shows the influence of the segmentation of the MEA. In this case the voltage of the cell with unsegmented MEA is clearly reduced resulting in less than 50% output power in comparison with the segmented MEA. As a reference the *VII* curve of three serial interconnected single cells with active area of each 0.18 cm^2 is added. The open circuit voltage

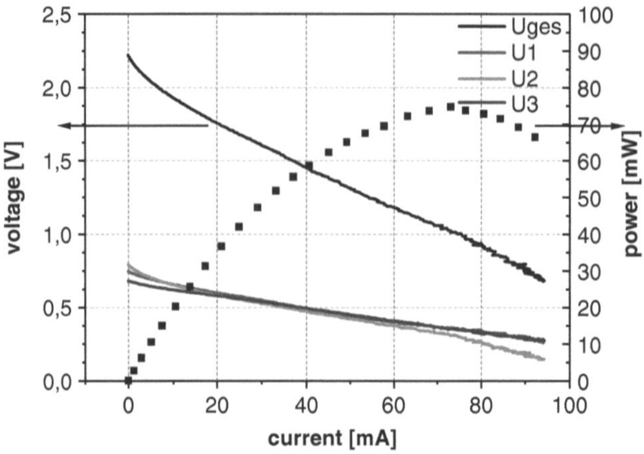

Figure 7-15. VII characteristic of micro fuel cell demonstrator, three cells 0.18 cm^2 serial interconnected, 25°C, 50% RH

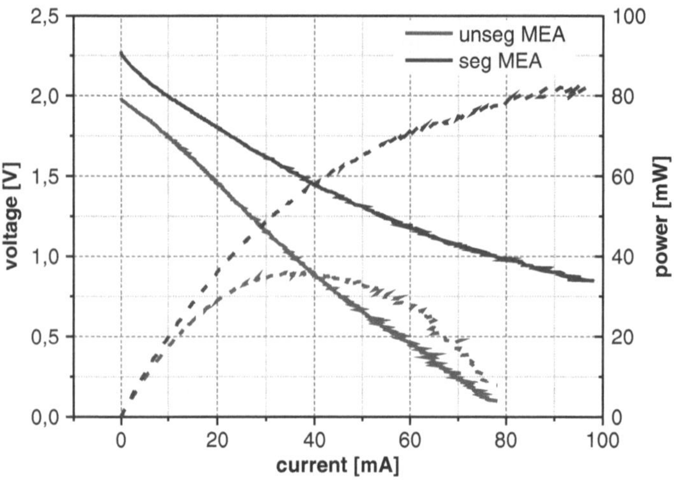

Figure 7-16. VII characteristic of micro fuel cell demonstrator with segmented MEA in comparison with an unsegmented MEA

of the reference cells is higher compared to the integrated and segmented micro fuel cell. But both curves merge already at ca. 10 mA and power density of both cells is almost identical. The lower voltage of the segmented MEA may be attributed to minor losses, probably due to internal leakage currents between adjacent cells.

An equivalent circuit of the segmented fuel cell is shown in Figure 7-17. R_L is the lateral surface resistance between neighbouring cells which may result from insufficient removing of the electrode layers in the segmentation procedure. V_{FC}, R_{ION} and R_{SI} are the open circuit voltage, the ionic membrane resistance and the resistance of the serial interconnection, respectively. To minimise internal losses

$$R_L >> R_{ION} + R_{SI} \quad (1)$$

must hold.

To prove the concept of leakage currents between cells through the electrode layers, the open circuit voltage of the non-patterned fuel cell of Figure 7-16 will be estimated.

At first the resistance R_L is calculated according to

$$R_L = \rho(d/wt) \quad (2)$$

with $\rho = 18.7\ \Omega$ mm, the specific resistivity of the electrode layer, $d = 200$ μm, the distance between adjacent current collector/flowfield of adjacent cells, $w = 7.6$ mm, the length of the cells and $t = 7.5$ μm, the thickness of the electrode layer, a resistance of $R_L = 66\ \Omega$ results.

As can be seen from Figure 7-17, at open circuit condition the outer fuel cells are shorted with one resistance R_L while the middle cell is shorted with two resistances R_L from both sides of the MEA.

The resulting open circuit voltage of the system V_0 can thus be calculated according to

$$V_0 = 2V_{FC}(R_L) + V_{FC} \times \frac{1}{2} R_L \quad (3)$$

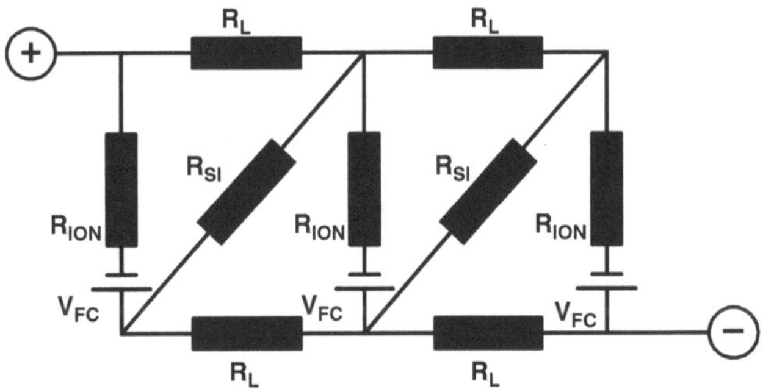

Figure 7-17. Equivalent circuit of the segmented fuel cell

with $V_{FC}(R_L) = 0.72$ V, the voltage of a single cell according to Figure 7-13 under the load R_L, $V_{FC}(R_L/2) = 0.67$ V, respectively. The resulting open circuit voltage is 2.11 V which is very close to the measured value of 1.98 V in Figure 7-16.

By comparing the *V/I* curve of the segmented MEA with the reference curve it can be seen in Figure 7-15 that the patterning process of 200 µm isolating lines can be performed without degradation of the MEA, and ionic drift currents between cells can be neglected.

Nevertheless, the total performance of the micro fuel cells is in the same range of current density compared to the best conventional cells and is at the same time highly miniaturised and capable for mass production which is a clear success of our micro-patterning approach.

CONCLUSIONS

Different technologies have been developed which allows the fabrication of PEM micro fuel cells based on commercially available MEAs. Planar micro fuel cells with size of 1 cm × 1 cm and 200 µm thickness with 40 mA output current at 1.5 V (=120 mW cm^{-2}, 25°C, 50% RH) has been successfully demonstrated. Cell performance was validated under varying ambient conditions. A peak power of 210 mW cm^{-2} was demonstrated with dry H_2 as fuel and with natural convection on the air side.

A maximum peak power density at optimal conditions of 210 mW cm^{-2} and stable long-term operation at 80 mW cm^{-2} was achieved at different ambient conditions. The total performance of the micro fuel cells is in the same range of current and power density compared to the best conventional planar PEM fuel cells. At the same time this technology offers a high degree of miniaturisation and the capability for mass production which is a clear success of our micro-patterning approach.

References

[1] R. Hahn and J. Müller, Future Power Supplies for Portable Electronics and Their Environmental Issues, International Congress ELECTRONICS GOES GREEN 2000+ (EGG2000), September 11–13, Berlin, Germany pp. 727–734.
[2] R. Hahn, M. Krumm, and H. Reichl, Thermal Management of Portable Micro Fuel Cell Stacks, Proceedings of the 19th IEEE Semiconductor Thermal Measurement and Management Symposium SEMI-THERM, San Jose, CA USA, March 11–13, pp. 2002–2009, 2003.
[3] K.B. Min, S. Tanaka, and M. Esashi, Silicon-Based Micro-Polymer Electrolyte Fuel Cells, Proceedings of the 16th International Conference on MEMS, Kyoto (Japan), pp. 379–382, 2003.

[4] K.-B. Min, S. Tanaka, and M. Esashi, Fabrication of novel MEMS-based polymer electrolyte fuel cell architectures with catalytic electrodes supported on porous SiO_2, *Journal of Micromechanics and Microengineering*, 16 (2006) 505–511.

[5] S.J. Lee, A. Chang-Chien, S.W. Cha, R. O'Hayre, Y.I. Park, Y. Saito, and F.B. Prinz, Design and fabrication of a micro fuel cell array with "flip-flop" interconnection, *Journal of Power Sources*, 112 (2002) 410–418.

[6] H.L. Maynard and J.P. Meyers, Proceedings of the 2nd Annual Small Fuel Cells and Battery Technologies for Portable Power Applications International Symposium, New Orleans (USA), 2000.

[7] J.P. Meyers and H.L. Maynard, Design considerations for miniaturized PEM fuel cells, *Journal of Power Sources*, 109 (2002) 76–88.

[8] J. Yu, P. Cheng, Z. Ma, and B. Yi, Fabrication of miniature silicon wafer fuel cells with improved performance, *Journal of Power Sources*, 124 (2003) 40–46.

[9] T.J. Yen, N. Fang, X. Zhang, G.Q. Lu, and C.Y. Wang, A micro direct methanol fuel cell operating at near room temperature, *Applied Physics Letters*, 83(19) (2003) 4056–4058.

[10] G.Q. Lu, C.Y. Wang, T.J. Yen, and X. Zhang, Development and characterization of a silicon-based micro direct methanol fuel cell, *Electrochimica Acta*, 49 (2004) 821–828.

[11] S. Aravamudhan, A.R.A. Rahman, and S. Bhansali, Porous silicon based orientation independent, self-priming micro direct ethanol fuel cell, *Sensors and Actuators A*, 123–124 (2005) 497–504.

[12] J. Yeom, G.Z. Mozsgai, B.R. Flachsbart, E.R. Choban, A. Asthana, M.A. Shannon, and P.J.A. Kenis, Microfabrication and characterization of a silicon-based millimeter scale, PEM fuel cell operating with hydrogen, methanol, or formic acid, *Sensors and Actuators B*, 107(2) (2005) 882–891.

[13] R.S. Jayashree, J.S. Spendelow, J. Yeom, C. Rastogi, M.A. Shannon, and P.J.A. Kenis, Characterization and application of electrodeposited Pt, Pt/Pd, and Pd catalyst structures for direct formic acid micro fuel cells, *Electrochimica Acta*, 50(24) (2005) 4674–4682.

[14] S.-C. Yao, X. Tang, C.-C. Hsieh, Y. Alyousef, M. Vladimer, G.K. Fedder, and C.H. Amon, Micro-electro-mechanical systems (MEMS)-based micro-scale direct methanol fuel cell development, *Energy*, 31 (2006) 636–649.

[15] D. Marsacq, Les micropiles à combustible, une nouvelle génération de microgénérateurs électrochimiques, *Clefs CEA* 50–51, 2004–2005.

[16] T. Pichonat, B. Gauthier-Manuel, and D. Hauden, A new proton-conducting porous silicon membrane for small fuel cells, *Chemical Engineering Journal*, 101(1–3) (2004) 107–111.

[17] T. Pichonat and B. Gauthier-Manuel, Development of porous silicon-based miniature fuel cells, *Journal of Micromechanical Microengineering*, 15 (2005) S179–S184.

[18] G.Q. Lu and C.Y. Wang, Development of micro direct methanol fuel cells for high power applications, *Journal of Power Sources* 144 (2005) 141–145.

[19] M. Müller, C. Müller, F. Gromball, M. Wölfle, and W. Menz, Micro-structured flow fields for small fuel cells, *Microsystem Technologies*, 9 (2003) 159–162.

[20] S. Ha, B. Adams, and R.J. Masel, A miniature air breathing direct formic acid fuel cell, *Journal of Power Sources*, 128 (2004) 119–124.

[21] S. Ha, Z. Dunbar, and R.I. Masel, Characterization of a high performing passive direct formic acid fuel cell, *Journal of Power Sources*, 158(1) (2006) 129–136.

[22] K. Shah, W.C. Shin, and R.S. Besser, Novel microfabrication approaches for directly patterning PEM fuel cell membranes, *Journal of Power Sources*, 123 (2003) 172–181.

[23] K. Shah, W.C. Shin, and R.S. Besser, A PDMS micro proton exchange membrane fuel cell by conventional and non-conventional microfabrication techniques, *Sensors and Actuators B*, 97 (2004) 157–167.

[24] S.H. Chan, N.-T. Nguyen, Z. Xia, and Z. Wu, Development of a polymeric micro fuel cell containing laser-micromachined flow channels, *Journal of Micromechanical Microengineering*, 15 (2005) 231–236.

[25] A. Schmitz, M. Tranitz, S. Wagner, R. Hahn, and C. Hebling, Planar self-breathing fuel cells, *Journal of Power Sources*, 5213 (2003) 1–10.

[26] Andreas Schmitz, Stefan Wagner, Robert Hahn, and Albrecht Weil, Segmentation of MEA by LASER Ablation, Fuel Cell Seminar, Luzern 2003.

Chapter 8

THERMAL CONSTRAINTS OF PEM MICRO FUEL CELLS FOR PORTABLE ELECTRONICS

Robert Hahn, Stefan Wagner, and Herbert Reichl
Fraunhofer Institute Reliability and Microintegration (Fraunhofer-IZM), Gustav-Meyer-Allee 25, 13355 Berlin, Germany

INTRODUCTION

Despite increasingly efficient components and low-power technologies, the energy demands of portable electronic products will rise dramatically in the future due to their growing functionality. As improvements in battery technology have so far been limited to energy density increases of only a few percent per annum, over the past few years many R&D activities have concentrated on alternative forms of portable power supply. One of the most promising candidates is micro fuel cells (FCs) based on polymer electrolyte membranes (PEMs) [1–3]. But there are several of challenges concerning fuel cells as a battery replacement:

- PEM FCs must breath air.
- PEM FCs tend to be more complex than batteries.
- PEM FCs require a uniform humidity in the membrane to have suitable ionic conditions.
- Cell temperature must be controlled. If the temperature of the MEA exceeds 80°C, thermal damage of the membrane may occur.
- Current Li-ion batteries achieve up to 500 W h l^{-1}. Thus the overall power density of the fuel cell system must be considerably higher.

This paper focuses on problems concerning the thermal management when using a fuel cell in a portable device. They can be summarized as follows:

- Heat dissipation is a performance limiting parameter in many portable electronic products.
- Fuel cell efficiency of ca. 50% increases total thermal load of a portable device by 80–100%.
- Thus it is a technical challenge that a system with portable fuel cell remains within allowable operating temperatures.

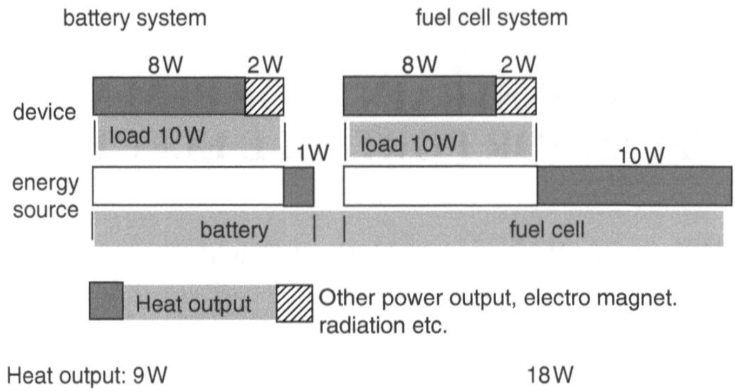

Figure 8-1. Comparison of thermal load of a battery and fuel cell powered electronic device

- A liquid cooling loop is not applicable in miniaturized systems.
- Accurate thermal models are needed to develop the systems integration of fuel cells.

As a demonstration, Figure 8-1 shows a comparison between of an electronic system powered by a battery and a fuel cell. The conversion efficiency of the battery and the fuel cell are set 90% and 50%, respectively. It is assumed that the device converts 80% of the electrical power into heat. The remaining 20% may be converted into electromagnetic radiation if a transceiver is incorporated, light radiation or others. The fuel cell system results in a twofold heat output compared to the battery system.

FUEL CELL PRINCIPLE

A fuel cell stack can be subdivided into three constituent component groups: the membrane electrode assemblies (MEAs) which fulfill the electrochemical function of the fuel cell, the bipolar plates (commonly consisting of graphite and carbon-filled polymers) which supply the MEAs with hydrogen and oxygen, providing cooling and conductive electronic paths, and the gas diffusion layers (GDL) which are inserted between MEAs and bipolar plates to distribute the reactants uniformly. The GDL consists of carbon fabric and Nafion. The fuel cell stack consists of a repeated interleaved structure of MEAs, GDLs, and bipolar plates. Current is produced across the surface of the MEAs and collected at the ends of the stack. The MEAs and multifunction plates are clamped together with a certain force, which is necessary to reduce electrical

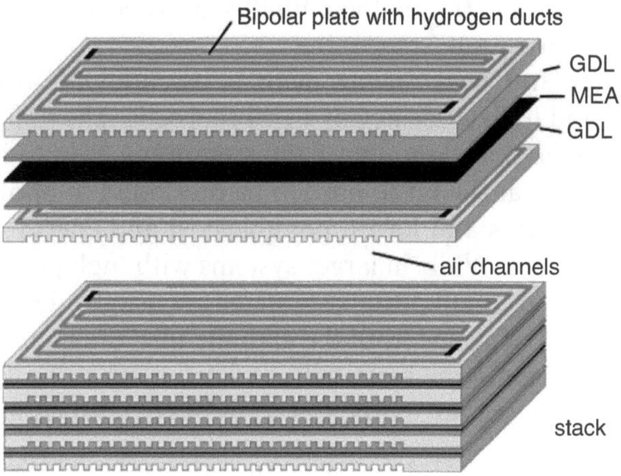

Figure 8-2. Basic construction of a PEM fuel cell stack

contact resistance between pressure assembled conductive components and to ensure the integrity of the reactant gas flow paths. A partial view of the stack under examination is given in Figure 8-2.

HEAT TRANSFER IN FUEL CELLS

The general energy balance equation for the energy accumulation in a fuel cell can be written as follows:

$$\mathrm{Sum}(m_i Cp_i \mathrm{d}T / \mathrm{d}t) = -I(\Delta H / nF + E_c) + \mathrm{Sum}(N_i A c_{pi}(T_{in} - T_{out}) - Q_{(-)}$$

$$E_c = E_o - \mathrm{Sum}\,(|\eta|) - IR_i,\ E_o = -\Delta G / nF \qquad (1)$$

were A and I are the total system area and the current flow, respectively, ΔH is the enthalpy change due to the fuel cell reaction, $\mathrm{Sum}(N_i A c_{pi}(T_{in} - T_{out})$ is the enthalpy change between the inlet and outlet for all components, N_i and m_i is the mass flow and mass of each component, E_c is the cell voltage, $Q_{(-)}$ is the enthalpy heat losses in the cell and E_o is the open circuit voltage.

$$Q_{(-)} = R_{vap} \Delta H_{vap} + KA(T - T_{amb}) \qquad (2)$$

The heat losses $Q_{(-)}$ are a result of the heat transfer through the walls of size A mainly by heat conduction, and the vaporization of water. K is the heat transfer coefficient. T is the fuel cell temperature and T_{amb} is the ambient temperature. ΔH_{vap} is the enthalpy of vaporization and R_{vap} the rate of water vaporization.

Proper water and heat management is essential for obtaining high power density and high efficiency of PEM fuel cells. The following aspects have to be considered: drying out of the membrane have to be prevented, water accumulation at the GDL have to be prevented to allow continuous access of the reactants and lowest possible cost and energy requirements of the cooling system. While in larger systems with high power density the hydrogen is hydrated, small portable systems are usually designed in a way, that back diffusion of water from the cathode into the membrane is sufficient to keep the membrane hydrated.

Principally there are three different ways of fuel cell cooling:

a. Active cooling with extra cooling channels in the bipolar plates. This can be realized with liquid cooling or air cooling
b. Cooling at the outer surfaces of the fuel cell stack
c. Design of a planar fuel cell and cooling only with natural air convection

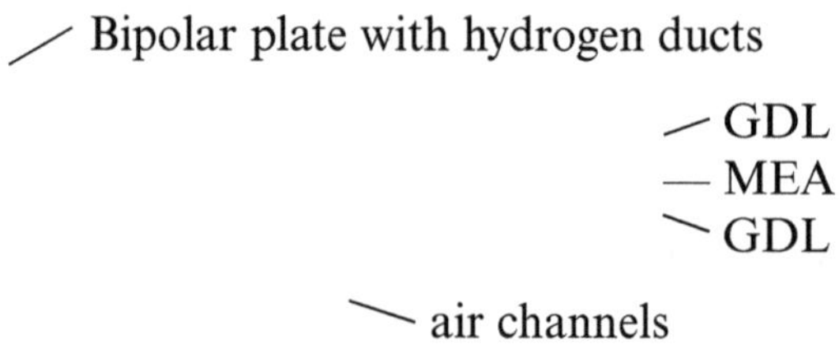

Due to cost and complexity issues only the variants *b* and *c* may be adequate for portable electronic equipment.

CHARACTERIZATION OF THE FUEL CELL STACK

A prototype of a PEM fuel cell system made of a stack of 15 bipolar plates, was developed to deliver 9 W output power at 8 V [4–6]. The stack of only 30 cm^3 volume (50 mm × 24 mm × 25 mm) was constructed using graphite micromachined bipolar flow-field plates. Adhesive bonding was used for assembly and sealing. The stack is shown in Figure 8-3.

In order to develop a suitable cooling method for the micro fuel cell stack a thermal model of the stack was created. The model was validated and unknown material parameters were fitted by means of thermal imaging experiments. Then different cooling concepts were tested with the aid of the thermal model. The work was done under the following constrains:

- No additional heating of critical components of the device may result from fuel cell integration.

Figure 8-3. Micro fuel cell stack

- Special cooling ducts in the stack-design are not possible for reasons of miniaturization and of economy.
- Cooling of the stack only over the cathode air supply is not sufficient.
- The cathode airflow has to be controlled within tight margins to maintain the membrane hydration state and is not sufficient for heat removal.

EXPERIMENTAL

The stack was tested in a computer-controlled measurement setup. Air was supplied to the cathode flow channels either by two fans or by means of a membrane pump and a flow controller to allow measurement of flow rate and pressure drop. Dry hydrogen was fed through a flow controller at a pressure of 300 mbar to the anode side. The temperature distribution of the stack was measured using a thermal imager[1] at the air outlet. High-performance water cold plates were mounted on the outer stack surface with thermal interface material acting as an isothermal boundary. Stack temperature was controlled using cold plates and a cryostat. Figure 8-4 shows the measurement setup.

Figure 8-5 shows the different cooling configurations used to study the thermal conductivity along and across the stack. The electrical load and *U*/*I* curves were measured using a battery test system.[2] The voltage of each cell was monitored separately. Figure 8-6 shows a typical *U*/*I* curve for the stack. Five fuel cell stacks were used for the experiments described here.

---

1 IR-M700 Mitsubishi.
2 Maccor series 4000.

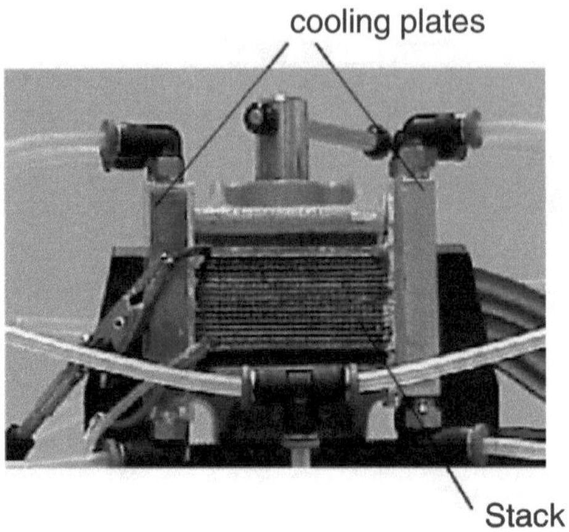

Figure 8-4. Experimental setup with lateral mounted cooling plates

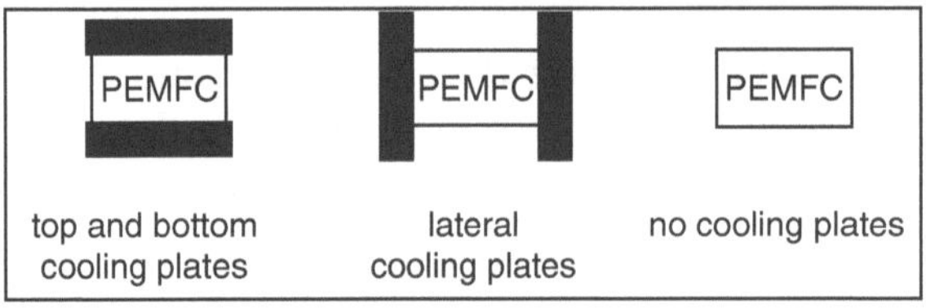

Figure 8-5. Cooling configurations used to develop the fuel cell stack thermal model (*Top* and *bottom* means, the cooling plates are mounted parallel to the both end plates of the stack. Lateral cooling means, the cooling plates are mounted perpendicular to the cells, contacting all the cells.)

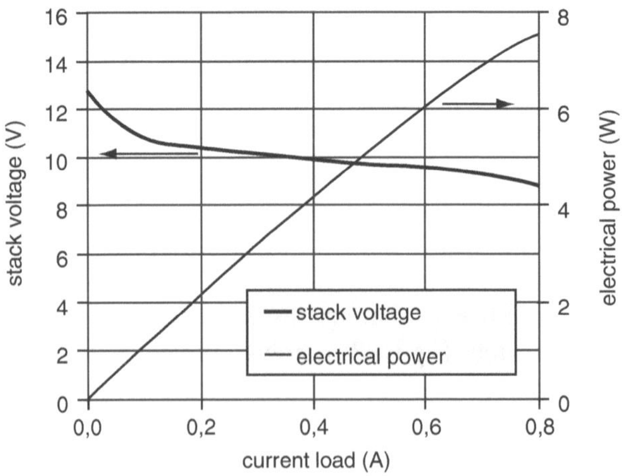

Figure 8-6. *U/I* curve of micro-PEM fuel cell stack

THERMAL MODEL

To design a cooling system for the fuel cell stack the thermal parameters of the stack and its components had to be determined accurately. For this reason a 3D FV (finite volume) thermal model was created.[3] Special emphasis was placed on the thermal conduction of the various layers responsible for heat flow in all directions of the fuel cell stack. The basic parameters are given in Table 8-1.

A homogeneous heat source was assumed at the location of the membrane electrode assembly. The heat generated was calculated according to (3).

$$P_v = (E^0_{rev} N I_L) - P_{el} \tag{3}$$

E^0_{rev} 1.23 V
N Number of cells
I_L Load current (mA)
P_{el} Produced electrical power (W)
P_v Power dissipation, waste heat (W)

The fan curves were measured and implemented in the model. Changing air parameters due to increased humidity resulting from the fuel cell cathode reaction were also taken into consideration. It was shown that the influence of the water on the temperature distribution is less than 1%.

The electrical connection between bipolar plates, GDL and MEAs is achieved by applying force between the endplates of the stack. Since there are variations in the electrical contact resistance as a function of mechanical pressure, with possible swelling of the MEAs, an analogous thermal contact resistance between the cells was introduced.

Two high efficiency liquid cooling plates were mounted at the top and bottom and at the left and right sides of the stack. In this manner, thermal conduction along and across the bipolar plates was investigated.

Table 8-1. Model parameters of PEM fuel cell stack

Component	Material	Dimensions $w \times l \times t$ (mm)	λ (W mK^{-1})
Bipolar plates	Graphite	50 × 25 × 1.4	Fitting parameter
Diffusion layer	Graphite/Nafion	50 × 25 × 0.19	1.676
Contact resist.	–	50 × 25 × 0.001	Fitting parameter
MEA	PRIMEA®5510	50 × 25 × 0.035	0.25
Active area of MEA		40 × 21 × 0.035	0.25
Current collector	Copper	50 × 25 × 0.2	394

[3] *FLOTHERM 3.2*, Flomerics Ltd.

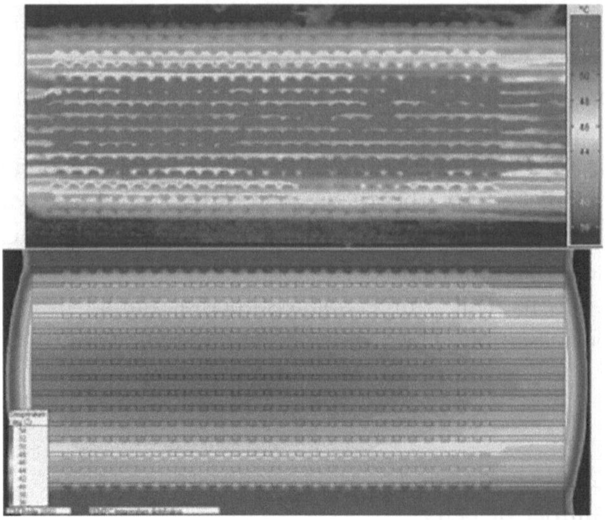

Figure 8-7. Comparison of measured (*top*) and simulated (*down*) stack temperature distribution

By comparing the temperature distribution of the measurement results with simulations for both configurations, the values for contact resistance and thermal conductivity of the graphite plates were fitted. Thus, a high model accuracy for all operating conditions was achieved. Deviations between model and measurements were below 0.5 K, as can be seen in Figures 8-7 and 8-8 for a power of 6.5 W and identical fit parameters. Parameters for the best match of simulation and measurements are outlined in Figure 8-9. The predicted contact thermal resistances between bipolar plate and gas diffuser layer and between gas diffuser layer and MEA were set the same (0.166 K W^{-1}) because they can not be distinguished by the curve fittings. As can be observed in Figure 8-8, cooling the stack from the sides is much more effective. In case of top and bottom cooling the temperature difference between inner and outer cell is 16°C at 6.5 W and can be even higher. This is not acceptable for stable long-term operation. It may be caused by the high thermal resistivity of the membranes and the thermal contact resistance between bipolar plates, GDLs and MEAs. Table 8-2 summarizes the results of temperature variations as function of operating and cooling conditions.

WATER BALANCE

The water balance of the membrane has to be taken into account when designing thermal management. PEMs require water to have suitable ionic conductivity. In the absence of sufficient hydration, the membrane

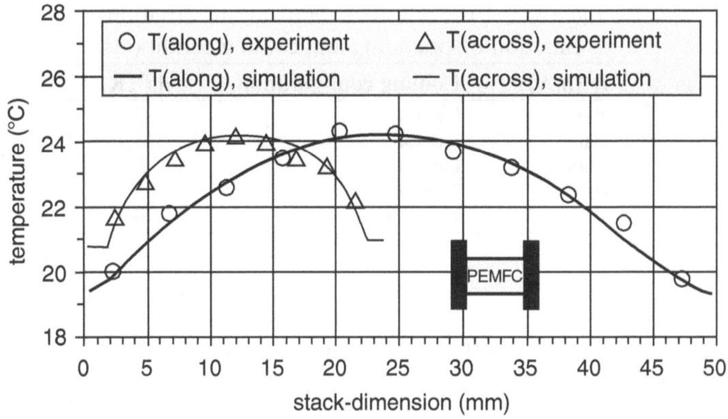

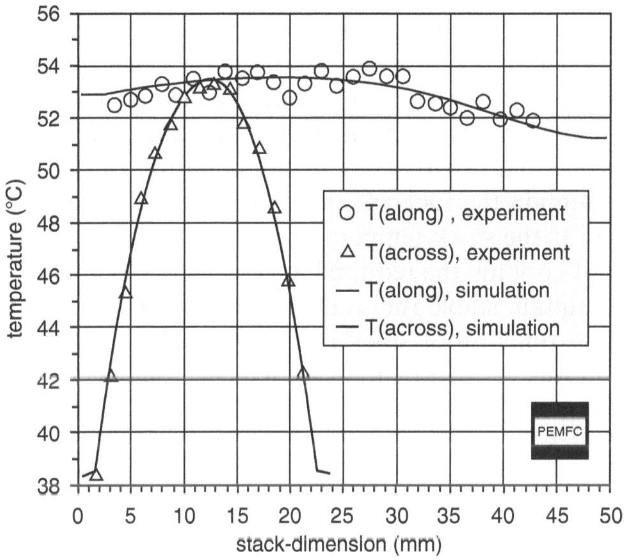

Figure 8-8. Comparison of measured and simulated temperature distribution for lateral (**a**) and *top/bottom* (**b**) cooling at different temperatures of the cold plates, I_L = 700 mA (along: temperature gradient of the middle bipolar plate; across: temperature gradient across all bipolar plates in the middle of the stack)

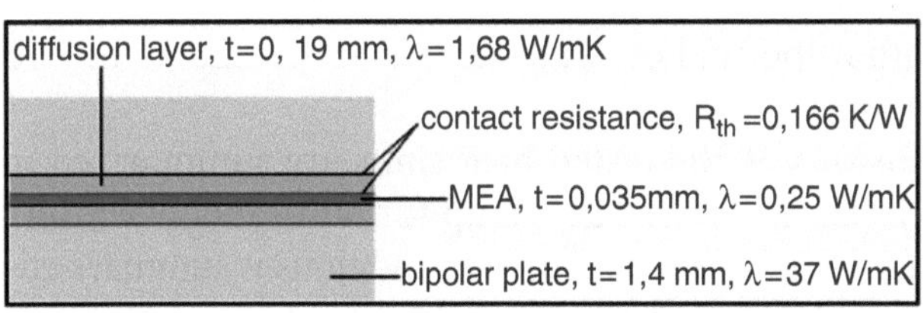

Figure 8-9. Resulting numerical fit parameters

Table 8-2. Fuel cell stack temperature variations as function of operating and cooling conditions

I_L (mA)	Cooling configuration	dT (K)
400	Lateral	2.5
700	Lateral	4.4
1,000	Lateral	7.4
400	Top/bottom	8.5
700	Top/bottom	16

becomes too dry and the ohmic drop across the membrane increases. At higher current densities, however, the water produced at the cathode can condense and form a liquid film that blocks oxygen transport to the cathode if airflow is insufficient. It was proven by simulation as well as in experiments that at temperatures below 60°C the stack can be operated up to a current of 200 mA with cooling only by means of the cathode stack air movers and natural convection at the sides. Since the air water uptake increases exponentially with rises in temperature, the airflow which is required to maintain the fuel cell membrane humid initially increases but then decreases. If the stack temperature is then kept constant by an additional means of cooling, the required airflow increases again with the load current. To maintain stable fuel cell operation the cathode airflow has to be controlled as function of stack temperature. The existence of a steady state condition was verified by long-term measurements for time intervals of several hours.

THERMAL MANAGEMENT OF SYSTEM

Only up to 10% of heat can be removed with the cathode airflow since the water balance must be maintained. To remove the waste heat from the system additional heat sinks are needed. It was established that heat can be transferred very efficiently at the stack sides. Commercially available heat sinks were mounted at the sides using electrically isolating thermal interface materials. To add as little extra weight and volume as possible to the system, high performance pin fin coolers were used. The heat sinks are made of pure aluminum with a thermal conductivity of above 220 W m^{-1} K^{-1} by means of impact extrusion.

The dimensions of the tested heat sinks are summarized in Table 8-3.

Figure 8-10 shows the model setup for natural convection studies. The maximum stack temperature at various conditions is summarized in Table 8-4. Cooler no. 2 works best in these conditions. At a heat power of ca. 6 W the stack reaches the maximum temperature of 80°C. For these calculations

Table 8-3. Heat sink dimension, base plate: 25 × 25 × 2 × mm³

No.	Number of pins	Layout	Pin height (mm)	Pin diameter (mm)
1	49	Inline	17	1.9
2	41	Staggered	20	1.9
3	49	Inline	23	1.9
4	41	Staggered	23	1.9
5	18	Staggered	23	2.9

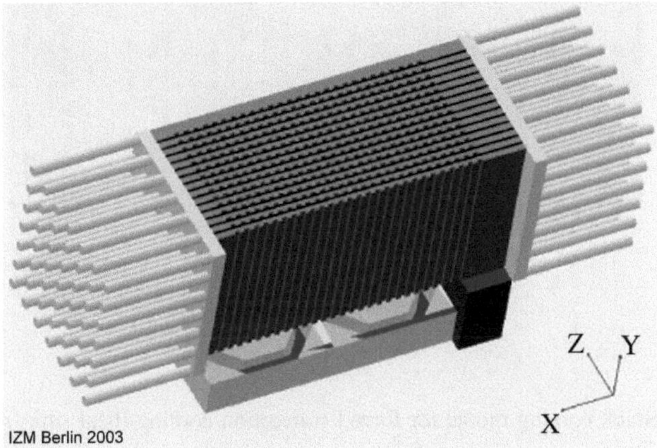

Figure 8-10. Fuel cell stack model with lateral heat sinks

Table 8-4. Maximum stack temperature with natural confection pin fin coolers at left and ride sides of the stack

I_L (mA)	P_v (W)	Number of heat sink	T_{max} (°C)
400	3.825	1	66.8
700	6.545	1	88.2
400	3.825	2	64.1
700	6.545	2	84.4

the airflow in the stack was adjusted according to the water management at the resulting temperature.

Space limitations or high ambient temperatures may impede the use of natural convection cooling. The heat sink performance can be improved and the size reduced by a factor 2–3 using forced convection.

Our intention was to use just one fan for both fuel cell stack cathode air supply and side cooling. The configuration applied is illustrated in Figure 8-11.

The coolers were mounted in flow direction on top of the fuel cell stack using extended base plates which make contact with the left and ride sides of the stack. Since airflow inside the stack has to be much lower than

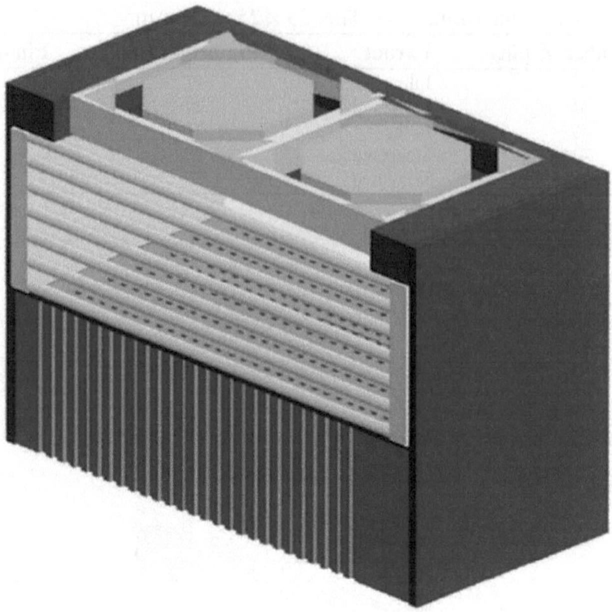

Figure 8-11. Stack cooling model for forced convection cooling (heat sink front side open only for visibility)

airflow in the pin fin region, a portion of the air leaving the heat sink had to be bypassed to the stack.

Figure 8-12 shows the impact of airflow. The temperature gradient of the extended heat sink base plate which transfers the heat from the stack to the heat sink as shown in Figure 8-13 accounts for a temperature rise of 8 and 14 K for Loads of 600 and 1,000 mA, respectively. This value can be reduced with further design improvements.

Figure 8-14 shows examples of the fuel cell microstacks used in the tests.

The first two rows of Table 8-5 represent experiments where the total airflow is passing through the fuel cell stack. In this case only low flow levels are applicable to prevent MEA dry-up as discussed above. The resulting stack temperatures are higher compared to natural convection (Table 8-4) since the air entering the stack has already been heated up.

The main conclusion from this study is that the waste heat can be transferred from the stack up to a full load of 8–10 W under forced air cooling conditions without exceeding 80°C maximum MEA temperature. One individual microfan can be used for both fuel cell cathode air supply and stack cooling. Table 8-6 compares both models in terms of overall performance. Since the fan consumes energy, the system efficiency reduces

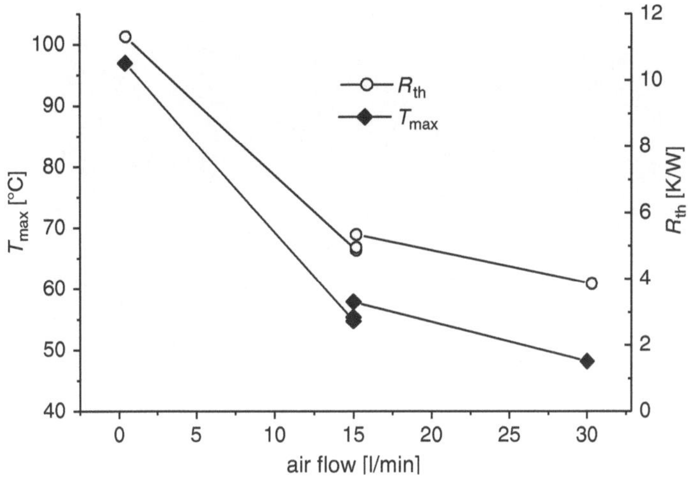

Figure 8-12. Influence of airflow on maximum stack temperature and R_{th}, I_L = 700 mA, model according to Figure 8-11

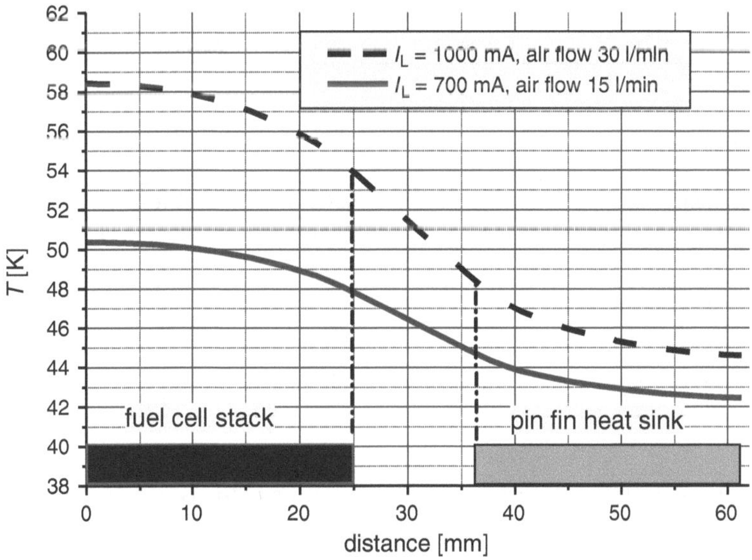

Figure 8-13. Temperature gradient at center line of heat sink base plate at I_L = 700 mA

Figure 8-14. Investigated micro fuel cells and pin fin heat sinks

Table 8-5. Numerical results of forced air pin fin cooling according to Figure 8-15

I_L (mA)	P_v (W)	Airflow (l min^{-1})	Number of heat sink	T_{max} (°C)
400	3.825	0.4	4	70.8
700	6.545	0.4	4	97
700	6.545	15	3	55.4
700	6.545	15	5	57.9
700	6.545	15	4	54.8
700	6.545	30	4	48.2
1,000	11.45	30	4	66.2

Table 8-6. Comparison of different fuel cell cooling methods with regard to systems efficiency

Air convection	Natural	Forced	Forced
Configuration	Figure 8-13	Figure 8-15	Figure 8-15
Number fans	1	1	2
Max. airflow (l min^{-1})	0.7	15	30
Max. power dissipation (W)	7.5	10	13
Fan power (mW)	60	200	400
$P_{el\,max}$ (W)	8	10	10
System efficiency % (stack + fan) at $P_{el\,max}$	49.6	49	48
System efficiency % at P_{el} = 4 W	49.2	47.5	45

with increasing fan current or number of fans. This is indicated in the last two rows of Table 8-5, where a peak power of 10 W and a stack efficiency of 50% are assumed. While efficiency in case of natural convection cooling is reduced only by 0.4–0.8%, it falls to as low as 45% in the case of two full-powered fans at P_{el} = 4 W.

FUEL CELL-POWERED CAMCORDER, NOTEBOOK COMPUTERS, AND MOBILE PHONES

Fuel Cell-Powered Camcorder

A prototype of a CamCorder equipped with a PEM fuel cell system made of a stack of 15 bipolar plates, as shown in Figure 8-15, was developed to deliver 9 W output power at 8 V. The stack described above was integrated into a housing which also includes the metal hydride reversible hydrogen storage, pressure control, fans, and electronics. This system replaces the Li-ion battery pack normally used as an energy source in the camera.

The PEM stacks in a configuration according to Figure 8-11 are positioned right underneath the viewfinder of the camcorder. Thus air inlet and outlet are on the left and right side and will not be blocked by the viewfinder which can be flipped and extended on top of the fuel cell system. The airflow is not disturbed by the user as well since the grip and controls are in front of the fuel cell. Therefore there is no thermal interference between camcorder and fuel cell system.

Fuel Cell-Powered Notebook Computers

The thermal design of a notebook computer is a difficult task and a variety of cooling methods have been employed [14,15]. To demonstrate

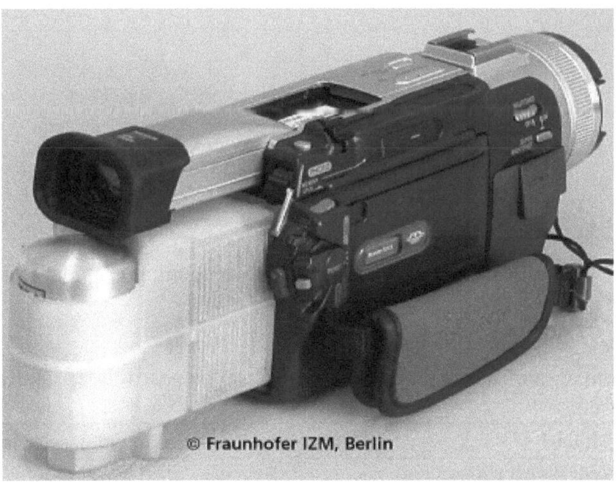

Figure 8-15. PEM fuel cell camcorder demonstration system

the impact of fuel cell integration into a notebook we analyzed a model with the following dimensions and boundary conditions:

Notebook size:	$28 \times 21 \times 2$ cm^3
Keyboard size:	26×10 cm^2
Total mean power:	24 W
Percentage of power used by display:	20%
Ambient temperature:	30°C
Maximum surface temperature:	40°C
Maximum surface temperature underneath keyboard:	60°C
Ambient air movement:	None

Figure 8-16 shows a typical load profile of the notebook under test.

Using these figures, we first estimated the maximum possible passive heat dissipation of the notebook housing. Following the calculations of [13], under such conditions, a maximum heat of 100 W m^{-2} can be dissipated from the notebook surface by natural convection and radiation, which equals 17 W in total for a notebook of these dimensions, as summarized in Table 8-7. Thus, at 24 W power a total of 7 W would have to be actively drained by fans, and it follows that the entire additional ca. 24

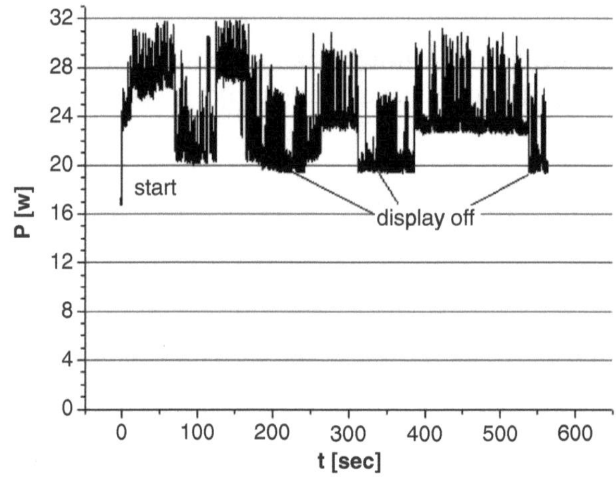

Figure 8-16. Electrical load of notebook computer under test

Table 8-7. Estimation of maximum power, which can be passively dissipated from notebook housing

Notebook constituent	Maximum passive heat dissipation
LCD panel, back side (in upright position)	6.7 W
Top and bottom chassis without key board	5.4 W
Key board	5 W
Sum	17.1 W

W produced by the fuel cell would also have to be actively removed by forced air convection.

State of the art heat sinks and miniature fans [16] capable of removing 24 W heat would occupy a volume of ca. 50 cm^3 with an additional energy demand of 1.3 W for active cooling or 100 cm^3 for passive heat sink cooling.

Using the stack geometry and stack cooling described in the previous section, the fuel cell stack occupies ca. 100 cm^3. The space requirement for the stack would therefore increase by 50–100% to fulfill the cooling requirements, so that the fuel cell stack and fan cooler would account for ca. 20% of the notebook's volume. Thus, even if the energy density of the stack can be increased by further miniaturization, the space necessary for heat removal will continue to pose a serious obstacle to miniaturization.

Fuel Cell-Powered Mobile Phone

In comparison with the notebook, today's mobile phones have a much lower energy density. Using the same approach as above, we estimated the passive heat dissipation potential of a mobile phone housing with the following dimensions and boundary conditions:

Total surface:	90 cm^2
Portion of surface with skin contact:	40 cm^2
Mean power during talk mode:	0.85 W
Mean power rf-radiation in talk mode:	0.25 W
Maximum surface temperature:	41.5°C

According to [13], approximately 220 W cm^{-2} can be dissipated by the human body with direct skin contact at a surface temperature of 41.5°C. Therefore, the maximum possible passive heat removal capability is: $0.005\ m^2 \times 100\ W\ cm^{-2} + 0.004\ m^2 \times 220\ W\ cm^{-2} = 1.4\ W$.

The mobile phone and fuel cell produce, respectively, 0.6 W + 0.85 W = 1.45 W of waste heat during talk mode. Taking into account that at least 5% of the fuel cell heat is carried out with cathode airflow, the surface of the mobile phone should be large enough for passive heat removal, provided that the heat can be spread evenly. Nevertheless, the device would need openings for supply and removal of the cathode air.

CONCLUSIONS

A 3D thermal model of a miniaturized PEM fuel cell stack was developed and validated under various operating conditions. It was used to study different cooling methods while maintaining the water balance of the system.

Based on these results the model was used to develop practical cooling constructions that allow the system integration of micro fuel cells into portable electronic products. They include thermal coupling of the stack with pin fin heat sinks. They are connected to the stack at the sides where thermal conductivity is at its highest. Natural convection cooling is adequate up to an electrical power of 8 W maintaining the highest system energy efficiency while forced convection can be deployed up to 10 W and higher ambient temperatures.

For electronic devices with low power density the additional heat of the PEM fuel cell can be passively dissipated with good housing conductivity and thermal coupling to the housing.

For electronic devices with high power density (notebook, etc.) active cooling (cooling elements and fans) are necessary which require an additional construction volume. If fans have to be avoided, passive cooling, or heat pipes can be applied which requires an additional construction volume.

Future studies will include transient thermal behavior, the influence of varying ambient temperatures and humidity as well as full systems integration.

ACKNOWLEDGMENT

The work was supported by the internal research programs of the Fraunhofer society.

References

[1] Global Portable Fuel Cells, Market Analysis & Forecasts, Strategic Research Report, Allied business, 2002.

[2] R.G. Hockaday, Microfuel Cells for Portable Electronics, Manhattan Scientifics, Power 2000, Sept. 25, 2000, Santiego.

[3] J.P. Meyers and H.L. Maynard, Design considerations for miniaturized PEM fuel cells, *Journal of Power Sources*, 109 (2002) 76–88.

[4] Robert Hahn, Michael Krumm, and Stefan Wagner, Thermal Management of Portable Fuel Cell Stacks, Fuel Cells Science & Technology 2002, Scientific Advances in Fuel Cell Systems, 25–26 September, 2002, Amsterdam.

[5] Robert Hahn, Michael Krumm, and Herbert Reichl, Thermal Management of Portable Micro Fuel Cell Stacks, Proceedings of the 19. IEEE Semiconductor Thermal Measurement and Management Symposium semi-Therm, San Jose, CA USA, March 11–13, 2003, pp. 2002–2009.

[6] DE10306081, Brennstoffzellenstack mit Bipolplatten, Hahn, Robert, 7.2.2003.

[7] H.I. Lee, T.Y. Oh, and I.W. Partk, Development of 1 kW class polymer electrolyte membrane fuel cell power generation system, *Journal of Power Sources*, 107 (2002) 110–119.

[8] R. Mosdale and S. Srinivasan, Analysis of water and thermal management in proton exchange membrane fuel cells. *Electrochemical Acta*, 40(4) (1995) 413–424.
[9] R. Kelley and S.D.J. Pratt, Method and apparatus for thermal management of fuel cell systems, Pat. US6406808.
[10] K.K. Sikka, K.E. Torrance, and C.U. Scholler, Heat Sinks with Fluted and Wavy Fins in Natural and Low-Velocity Forced Convection, ITHERM 2000, May 23–26, Las Vegas, NV, USA.
[11] H.H. Jung and J.G. Maveety, Pin-Fin Heat Sink Modeling and Characterization, IEEE 2000, March 21–23, San Jose, CA, USA.
[12] H.C. Chien, M.H. Tseng, and C.Y. Wang, The Study of Micro-Fin Heat Sinks for Electronic Cooling Applications, IEEE 2001, March 20–22, San Jose, CA, USA.
[13] T. Starner and Y. Maguire, A Heat Dissipation Tutorial for Wearable Computers, MIT Media Lab., MA 02139.
[14] Takashi Kobayashi, Tetsuro Ogushi, Noriaki Sumi, and Masao Fujii; Thermal Design of an Ultraslim Notebook Computer, IEEE Transactions on Components and Packaging Technologies, Vol. 23, No. 1, March 2000.
[15] J.R. Rujano, R. Cardenas, M.M. Rahman, and W.A. Moreno, Development of a Thermal Management Solution for a Ruggedized Pentium Based Notebook Computer, Thermal and Thermomechanical Phenomena in Electronic Systems, 1998. ITHERM '98. Seattle, 27–30 May 1998.
[16] C.K. Loh, D. Nelson, and D.J. Chou, Thermal Characterization of Fan-Heat Sink Systems in Miniature Axial Fan and Micro Blower Airflow, 17. IEEE Semi-Therm Symposium, 2001, 0-7803-6649-2.

Chapter 9

A DIRECT METHANOL FUEL CELL USING CERMET ELECTRODES IN LOW TEMPERATURE COFIRE CERAMICS

W. Kinzy Jones, Naveen Savaram, and Norman Munroe
Department of Mechanical and Materials Engineering,
Florida International University, Miami, FL 33199 Jones@fiu.edu

INTRODUCTION

Low temperature cofired ceramics (LTCC) is becoming more integrated as ceramic microsystems. Microsystems allow the integration of numerous components, including embedded passives, high-density interconnects, high-performance thermal management systems, sensors and actuators, mechanical, fluidic, and optical components. The development of these systems have led to enhanced processing capabilities including enhanced properties by controlled sintering or the development of cavities and microelectromechanical structures using fugitive inserts which are removed during firing. LTCC devices have produced meso and macroscale channels, large volume cavities, microcavities, wick, and controlled porosity structures. LTCC has been used to fabricate the cavities and channels required in a direct methanol fuel cell (DMFC), using an externally attached membrane electrode assembly (MEA) which consists of a carbon matte anode with a platinum catalyst, a Nafion® proton exchange membrane (PEM) and a platinum/ruthenium cathode. The DMFC has a maximum thermodynamic voltage of 1.18 V at 25°C, defined by the anode and cathode half cell reactions:

Anode reaction: $CH_3OH + H_2O = CO_2 + 6H^+ + 6e^-$

$E_a = 0.046$ V

Cathode reaction: $3/2O_2 + 6H^+ + 6e^- = 3H_2O$

$E_c = 1.23$V

Cell reaction: $CH_3OH + H_2O + 3/2O_2 = CO_2 + 3H_2O$

$E_{DMFC} = 1.18$ V

However, the cell voltage is much less than this, typically 0.3–0.4 V due to the losses from poor cathode activity, catalyst loading, and impedance sources within the cell.

In the above reactions, the oxidation process takes place in the anode electrode where the methanol is oxidized to carbon dioxide, protons, and electrons. In the reduction process, the protons combine with oxygen to form water and the electrons are transferred to produce the power. Figure 9-1 is a reaction scheme describing the probable methanol electrooxidation process (steps i–viii) within a DMFC anode [1]. Only Pt-based electrocatalysts show the necessary reactivity and stability in the acidic environment of the DMFC to be of practical use [2]. This is the complete explanation of the anodic reactions at the anode electrode. The electrodes perform well due to the presence of a ruthenium catalyst added to the platinum anode (electrode). Addition of ruthenium catalyst enhances the reactivity of methanol in fuel cell at lower temperatures [3]. The ruthenium catalyst oxidizes carbon monoxide to carbon dioxide, which in return helps methanol reactivity with platinum at lower temperatures [4]. Because of this conversion, carbon dioxide is present in greater quantity around the anode electrode [5].

The increase of the reactivity of methanol in DMFC is due to the breaking of carbon-to-carbon bond (C–C bond).

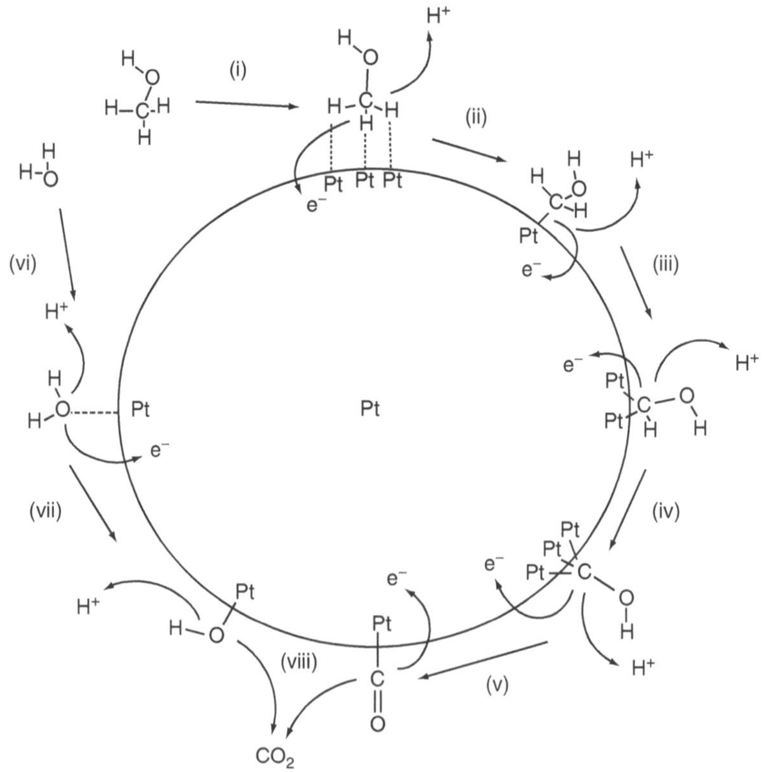

Figure 9-1. Schematic of methanol electrooxidation in the anode of DMFC [1]

In this work, a cermet electrode structure is developed for the anode structure using a porous borosilicate structure as the mechanical base for the anode structure coated with a thin silver film from the reduction of Ag resinate film to provide a conductive base. The electrode catalysts were added by the reduction of a mixture of platinum/ruthenium resinate.

Although this work only evaluated the performance of the anode electrode, a complete fuel cell is envisioned that would be cofired into the LTCC, consisting of the required cavities for fuel storage, channels, and wick structures for fuel deliver, a porous Ag structure for the electrodes, separated by a cavity with nanoporous surfaces that will allow the PEM organics to be added after firing and polymerized in situ. This work only considers the development of the porous cermet films for the Ag-based anode structure.

REVIEW OF DIRECT METHANOL FUEL CELL ELECTRODES AND CATALYST

Prior work has evaluated parameters that effect fuel cell performance. Researchers have used different types of catalysts (Pt, Ru, Se, etc.) with various combinations of catalysts ratios, methanol concentrations, operating temperatures, and cell design. Since the catalyst is a major cost factor, research has focused on design which minimizes the catalyst content and manufacturing cost, while maximizing performance.

For a general overview of fuel cell technology, Lunghi and Ubertini [6] have reviewed the procedures and parameters for single cell performance evaluation for the first time in fuel cell technology. The first part of the work was written so nontechnical personnel could understand better with minimum technical knowledge of fuel cells [7]. The last part of work was focused for technical personnel who have knowledge of some fuel cell technology but are not completely aware of all the parameters of other fuel cells. This effect tried to produce a methodology for evaluation of performance of all fuel cells on a common ground, which was not possible before. The parameters discussed were: fuel cell type molten carbonate fuel cell (MCFC), solid oxide fuel cell (SOFC), alcohol fuel cell (AFC), DMFC, polymer electrolyte fuel cell (PEM)), materials (for electrodes and electrolyte), age, operating temperature, operating pressure, current, anode inlet flow rates, cathode inlet flow rates, anode gas utilization factor, cathode utilization factor, dilution flow rates at both anode and cathode, steam flow rate, contaminant concentration, and active area.

Arico et al. [8] gave a complete view of DMFC from construction to futuristic development. The types of catalysts used in carbon electrode, types of membrane used, and types of testing done on the basis of various compositions of catalyst [9] and methanol are discussed in detail. With high surface area of Pt–Ru catalysts, the specific activity increases with the particle size. The particle with a growth habit in the plane (111) plane has high-surface area which increases the performance of the fuel cell.

Xin et al. [10] evaluated the synthesis of catalyst nanoparticles using microemulsions like cyclohexane. In this work no pure metal were observed, only alloys of the catalyst exhibited as seen from X-ray diffraction (XRD). The composition and particle was controlled using microemulsions [11,12]. The alloy formation of platinum, which has a FCC structure, included shifting of the diffraction peak to a higher angle (2θ), indicating a decrease in lattice parameter.

Vigier et al. [13] used three types of catalysts (platinum, ruthenium, and tin) in a DMFC. The atomic compositions of the catalysts were varied to get the best performance from each of the combinations. Under different voltammetric conditions Pt–Sn, electrocatalysts are the most active combination toward the methanol oxidation and production of CO_2, compared to Pt and Pt–Ru electrodes [14–16].

Coutanceau et al. [17] reveals the preparatory techniques for Pt–Ru bimetallic anodes by galvanostatic pulse electrodeposition in DMFC. The preparatory techniques are focused on reducing the noble metal catalyst loading of the electrodes and associated cost without decreasing the efficiency. To increase the performance of the electrodes [18,19] (i.e., the true surface area of the catalyst) one needs either to increase the thickness of the active layer, for a given catalyst loading; or to increase the amount of catalyst in the catalytic powder. Increasing the thickness of the active layer leads to a decrease in the diffusion rate of the reactant toward the catalytic sites, whereas increasing the weight loading generally leads to an increase in the particle size of the catalysts, thus decreasing their efficiency.

The influence of structure and composition on the reactivity of Pt–Ru catalysts for anodic methanol oxidation was studied by Hoster et al. [20]. Many surface reactions require suitable atomic configurations at the surface, which allow the reactants to be in an appropriate binding situation and geometric orientation towards each other. The preparatory methods for catalysts allow only limited control over the resulting surface structure, and generally the Ru concentration rather than its geometric distribution may be controlled [21–26].

Takasu et al. [27] prepared a homogenized Pt–Ru/C electrocatalyst with a high-specific activity for methanol oxidation from carbon black and ethanolic solutions of $Pt(NH_3)_2(NO_2)_2$ and $RuNO(NO_3)_x$. The specific activity for methanol electrooxidation increased with an increase in the Pt/Ru particle size. The concept of larger particle size aiding in the activity of methanol oxidation was experimentally verified [28–33].

Lizcano-Valbuena et al. [34] have investigated that the particle size increases as a consequence of the thermal treatment which aids in the performance of DMFC. This was achieved by the reduction of Pt–Ru/C catalysts with formic acid. Different flow fields were also tested as a part of the performance analysis. Pt–Ru composition ratio was 75:25 with an optimum operating temperature at 90°C.

Lim and Wang [3] reported that catalyst amount was reduced to 4 mg cm^{-2} in the anode electrode and 1.3 mg cm^{-2} in the cathode electrode of DMFC. The power generated with the above catalyst loading in the electrodes is 0.21 W cm^{-2} with high temperature and high-methanol concentration. The power generated was under the operating conditions of nonpressurized anode side and nonhumidified air pressurized to 15 psi. The initial power generated by the DMFC was 0.1 W cm^{-2} [40] which is half of the power generated after the optimization of the electrodes. The carbon fiber used as the conducting material had hydrophilic characteristics which promoted the holding of water for the complete period of the methanol reaction.

Min et al. [35] experimented on high-catalyst loading with 60% carbon and 40% Teflon backing claimed to be the most efficient electrode for direct methanol/proton exchange membrane fuel cell (PEMFC). The catalysts used were platinum and ruthenium which formed an alloy at an atomic ratio 1:1. The formation of the alloy was seen in XRD as there were no pure metal peaks found. The alloy formation of Pt and Ru promotes oxidation of methanol at lower temperatures. The 60% carbon backing makes it evident that the lower the percentage of carbon increases the efficiency.

Neerat et al. [4] evaluated the variations of weight loadings of catalyst (Pt, Ru) in the electrode of DMFCs. The various catalysts loading were tested in a half cell arrangement in sulfuric acid at various temperatures. Upon optimization, a 60% carbon supported electrode, Pt–Ru catalyst showed maximum mass activity. Higher activity was reported above 60°C when the ration of Pt to Ru is 1:1 rather than 3:2. At lower temperatures the 3:2 ratio of Pt to Ru exhibited higher activity. A 60% carbon electrode proved that elimination of carbon from the DMFC system is the best option for the improving its efficiency [36].

Lamy et al. [37] have done extensive research on methanol adsorption and oxidation on fuel cell through radioactive labeling study and FTIR measurements. Kinetics of formation of surface and bulk products coming from methanol and surface/bulk exchange processes were found to be significantly different on Pt compared to Pt/Ru [38,39]. Modifications of the electronic surface of platinum atoms by ruthenium were clearly seen in the electrode structure. Adsorption of carbon monoxide from methanol was reported as a very slow process at lower temperatures. The lowering of the operating temperature reduces the size of the unit as well as the start-up time of fuel cell [40,41]. Ruthenium increases the tolerance of anode surface poisoning [42,43]. This is complete study about how to control the poisoning of the anode of DMFC.

Scott et al. [44] have designed two types of flow cells (1) parallel flow channel arrangement and (2) a spot design of a flow bed. These two types of flow cells are designed to test with electrodes made of Teflon and carbon with ruthenium and platinum as catalysts on electrodes. Teflon was used for the purpose of providing hydrophobic effect in the electrode. By making the electrode hydrophobic, the flow of methanol is improved.

The research of Mallouk and Smotkin [45] considered combinatorial catalyst development methods. In the combinatorial research, the tools of electrochemical analysis (steady-state and dynamic voltammetry, chronoamperometry, scanning electrochemical microscopy, spectroelectrochemistry, complex impedance analysis) are used to test electrochemical cell. These tools allow the kinetic and mechanistic studies not readily available in nanoelectrochemistry. The research concentrated on improving the metallic catalyst, and also optimizing the interfacial contact and utilization.

Arrigo et al. [46] reported preliminary results concerning the micromachining procedure for fabricating a Si-based electrocatalytic membrane for miniaturized Si-based PEMFCs. Conventional batteries are already unable to deliver power in smaller volumes within the constraint requirements of long duration and light weight. A possible solution to overcome these limits is the use of a miniaturized fuel cell [47]. The next step to miniaturization is LTCC technology, where the complete fuel cell is manufactured in ceramic tape as demonstrated in this work.

Lu and Wang [48] have reported that by using stainless steel plates as bipolar plates (electrodes) with the flow field machined with photochemical etching technology; a micro-DMFC produce maximum power density of 62.5 mW cm^{-2} at 40°C and 100 mW cm^{-2} at 60°C at atmospheric pressure. The active electrode area is only 1.625 cm^2. The power generated by the stainless steel DMFC is double to that of the Si-based DMFC which the authors had previously fabricated.

The overall view of the literature is that the DMFC technology is progressing toward miniaturization and possibly replacing the carbon electrode, recognized as a depleting power factor due to CO poisoning of Pt catalyst or by decreasing the percentage of carbon in the process of developing the electrode. This gives rise to a promising LTCC technology which serves the purpose by replacing the carbon completely with ceramic glass/silver catalyst electrodes.

EXPERIMENTAL

The borosilicate electrode is a circular disc of 7 mm diameter and 50 μm thickness obtained from Robu glass from Andrews Glass Company (Germany). The circular disc is cut to approximate thickness using a diamond saw and then ground to the required thickness. The substrate is first coated with Ag resinate and fired at 360°C. Metallorganics are molecules with a single metal ion incorporated in their structures. On firing, the organics decompose, leaving a film of the metal, in this case silver. The amount of silver in the metallorganic ink is 25% [49]. Silver metallorganic is conductive, based on the theory of thin film conductivity. The substrate is spin coated with silver ink and vacuum impregnated followed by drying at room temperature for 2 h. The substrate is then fired with a firing cycle (1°C min^{-1} until it reaches 360°C). For uniform distribution of catalysts, the Pt and Ru resinates are mixed in an organic thinner, cyclohexane, and vacuum impregnated into the Ag-coated borosilicate substrate. The amount of metal present in the resinate is 14 and 4.5%, respectively, for the platinum and ruthenium resinate. As a conducting media silver resinate is used on the electrode. Organic solvent of the metallorganic catalyst was changed by adding additional solvent which would help in the uniform distribution of catalyst and also for optimizing the amount of catalyst loaded in the electrode.

The platinum/ruthenium mixture is then fired with the same firing cycle. The ratio of platinum to ruthenium is 1:1 [2,50] with the effective concentration, based on the flat area of the disc, applied to the substrate being 4 mg cm^{-2} [35,51]. The voltage–discharge curves of a commercially available fuel cell (H-Tec™ Methanol Fuel Cell Junior) was characterized. The cell was disassembled and the Ag-coated borosilicate anode was inserted and retested.

RESULTS AND DISCUSSION

Figure 9-2 shows the discharge curve of borosilicate electrode versus the standard carbon electrode of the H-Tec™ fuel cell. The load across the fuel cell for this test is 10 Ω, resulting in a discharge of 100 mA cm^{-2}. The cermet electrode demonstrates minimal increases of impedance over the discharge period and the higher overall voltage. The maximum developed by borosilicate substrate is 0.3489 V. This demonstrates that Ag metallization with a platinum/ruthenium catalyst can be developed as a cathode structure in DMFCs.

Cyclic voltammetry and Tafel plots were conducted to understand the electrochemical performance of the electrode. The Tafel Extrapolation

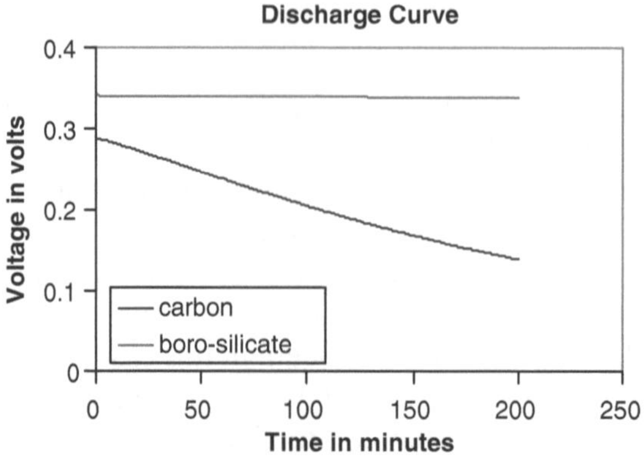

Figure 9-2. Discharge curves for cermet anode versus standard carbon supported anode at 100 mA cm^{-2} load

plot is a method for determining corrosion current density (corrosion rate) from polarization curves that characterize the kinetics of a corroding electrode with both anodic and cathodic reactions under activation control. The corrosion consists of two opposing electrochemical reactions: (1) An anodic reaction, in which a metal is oxidized and releases electrons via an external circuit; (2) a cathodic reaction, in which a solution species (often O_2 or H^+) is reduced, by receiving the electrons released by the metal.

The sharp point in the Tafel curve corresponds to the corrosion potential and the point where the extrapolated straight lines intersect correspond to the log of the corrosion current. In Figure 9-3 the Tafel plot of the Ag electrode exhibited a sharp point at approximately 0.1 V, at a current density of approximately $10^{-5.25}$ A/cm^2.

When the above Figures 9-3 and 9-4 are compared the carbon electrode discharges at a slow rate and the corrosion potential (sharp point) occurs at –0.15 V and a current density of approximately 10^{-6} A/cm^2. The performance of the DMFC extends from the start up time to the time needed to reach the maximum voltage and the duration of time the power is generated. The discharge curve and Tafel plots indicate that the ceramic electrode out performs the carbon electrode as the start-up time, the time to reach open circuit voltage after the addition of fuel, has been observed to be 50% shorter in the Ag based electrode. The current under load is rapidly established and reaches the stable condition much sooner than with the carbon electrode. The current density of the ceramic electrode was 1.3×10^{-5} Amp cm^{-2}, while the carbon electrode was 2.9×10^{-6} Amp cm^{-2}. With a higher current density, the power generation is greater and the fuel cell reaches stability faster.

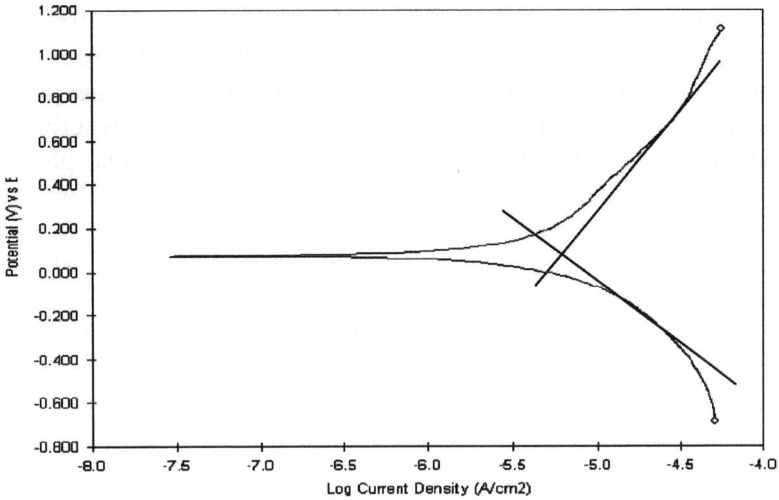

Figure 9-3. Tafel plot of Ag cermet electrode

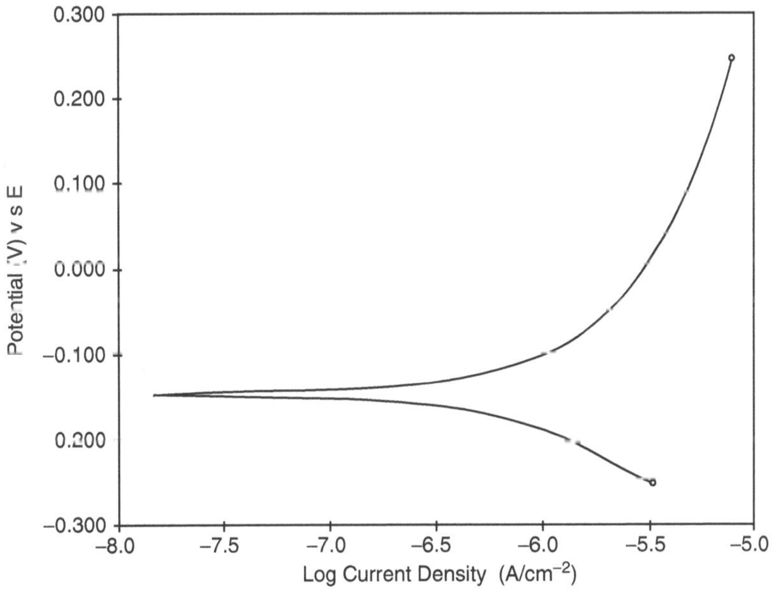

Figure 9-4. Tafel plot of carbon electrode

Cyclic Voltammetric Test

Cyclic voltammetry (CV) is an electrolytic method that uses microelectrodes and an unstirred solution where I/V curves (voltammograms) are obtained by sweeping the potential from the open circuit potential (E_{OC})

towards negative potentials. A peak produced in this step is attributed to a reduction reaction. When the potential is initially swept from the E_{OC} towards the positive direction, the anode current presents an exponential dependence on potential and corresponds to the oxidation reaction.

The cyclic voltammetry analysis of carbon and ceramic glass electrodes illustrated that the area formed in the swept V-I curve for the carbon electrode is less than ceramic electrode. Figures 9-5 and 9-6 illustrate the output voltage (E_{OC}) generated by the ceramic electrode is 294 mV and the carbon electrode is –89.8 mV. The areas formed across the current and voltage parameters define the power generated by the electrodes. The potential range of both the carbon and ceramic electrodes were kept constant to check the current variation. The current generated by the ceramic electrode was in the range of –120 μ amp to 120 μ amp and carbon electrode was –40 μ to 60 μ amp (for this particular fuel cell). The range of the current generated by the electrodes is proportional to the power generative performance of each electrode. The operating voltage of the ceramic electrode was in the range of pure platinum [52], ruthenium compound [5] and

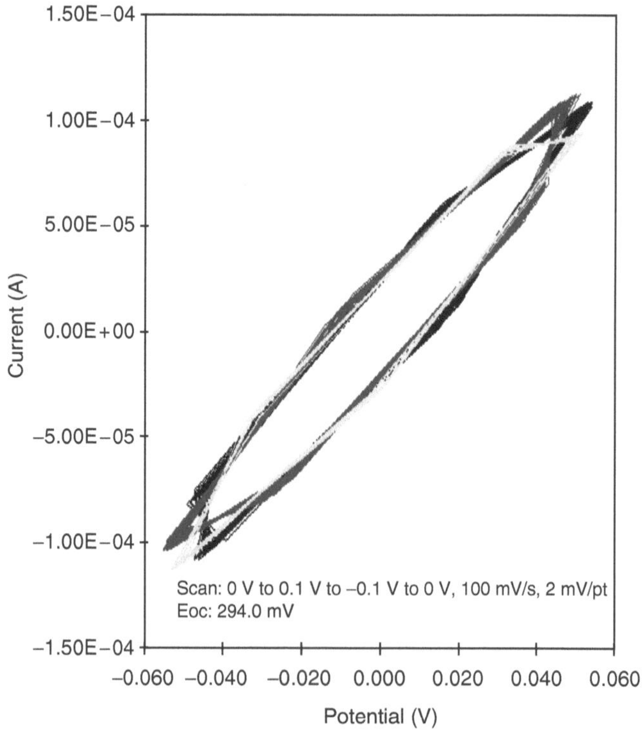

Figure 9-5. Cyclic voltammetry of cermet electrode

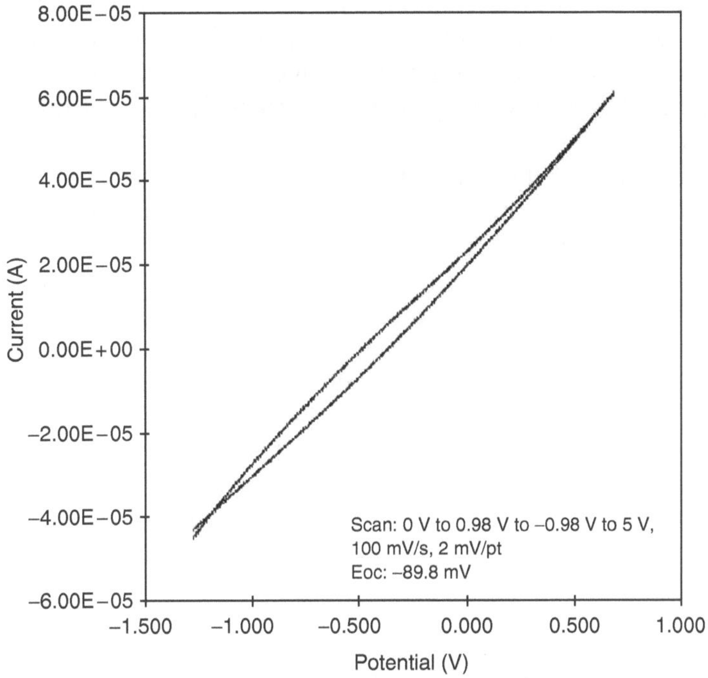

Figure 9-6. Cyclic voltametry of carbon electrode

Table 9-1. Voltage and current ranges in the cyclic voltameteric tests

Type of electrode	Voltage range (V)	Current range
Cermet electrode	–0.6 to 0.6	–120 to 120 μA
Pure platinum electrode	–0.2 to 0.8	Not available
Ruthenium electrode	0 to 0.8	–1 to 0.75 mA
Carbon electrode (Pt and Ru)	0 to 0.8	–0.2 to 0.05

that of carbon supported platinum ruthenium electrode [33]. The operating voltage and current ranges are shown in Table 9-1.

The working voltage range of Pt wire and Pt/GC (graphite carbon) electrodes were –0.2 to 0.8 V where as that for the ceramic electrode was –0.6 to 0.6 V, a wider range due to higher methanol reactivity and low start up time. The ruthenium electrode was operated at 0 to 0.8 V and –1 to 0.75 mA, the working range of ceramic electrode.

Cyclic voltammetry was conducted in methanol and non-methanol media with the former media producing higher operating ranges [5]. The carbon electrode with platinum and ruthenium catalysts (50:50 alloy) operated at 0 to 0.8 V and –0.2 to 0.05 A [33], which was also in the operating range of the ceramic electrode. This was due to the higher rate of methanol oxidation and reduction based on numerous factors (e.g., morphology, structure, particle size, composition, impurity, uniformity or dispersion of Pt-Ru catalysts). The rate of reaction of the ceramic electrode was faster than rates produced with pure platinum metal, ruthenium compound and platinum, ruthenium with carbon electrodes.

Figure 9-7a,b is SEM images of the cermet electrode at high and low magnification.

The SEM images show that the ceramic electrode has a uniform structure which aids in the uniform distribution of catalyst. The ceramic electrode demonstrates both a mesoscale porosity that assists in fluid transport, while the mesoscale particles have a nanoscale porosity which forms the base for the uniform distribution of the catalyst. The nanoporous structure demonstrates neck sintering which maximizes the surface area, assisting in the distribution of catalyst and higher methanol transportation from enhanced wicking. Transmission electron microscopy (TEM) was performed to understand the morphology and chemical analysis of the electrode structure. Figure 9-8 shows the amorphous structure of the glass/ceramic particle with transparent region at the neck sintering areas connecting the particles. TEM analysis demonstrates a microstructure of Pt–Ru catalysts

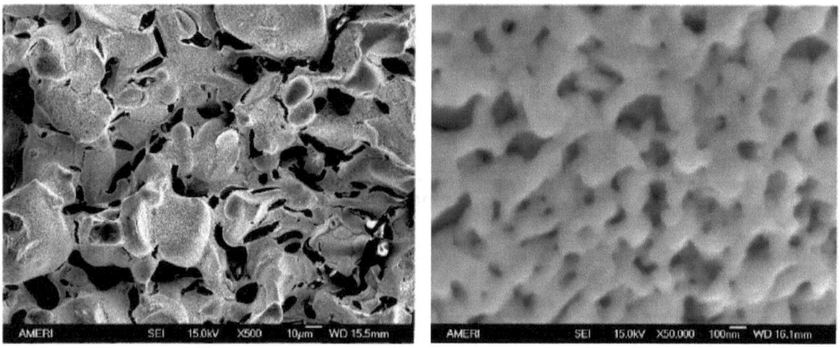

Figure 9-7. Low (500×) and high magnification (50,000×) of ceramic substrate

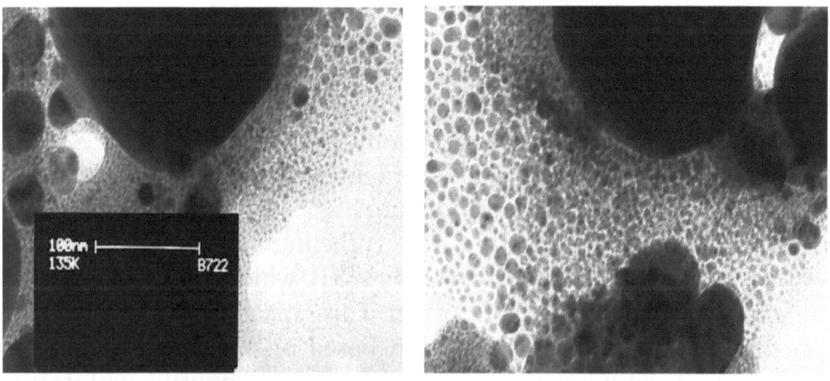

Figure 9-8. TEM micrograph of electrode structure

and Ag conductive media with a particle size below 20 nm. To understand the elements present, electron diffraction analysis was done.

The crystal structure of Ag and Pt are FCC while ruthenium is HCP. Additionally, ruthenium forms a solid solution in platinum in excess of 60 atomic percent platinum (Figure 9-10). The amount of silver used was almost 20 times more than that of the catalysts. The selected area electron diffraction analysis was performed on the electrodes, demonstrating both a ring pattern of silver and spot patterns indicative of platinum (Figure 9-9). No individual spots/rings were observed for ruthenium, as expected from the phase diagram.

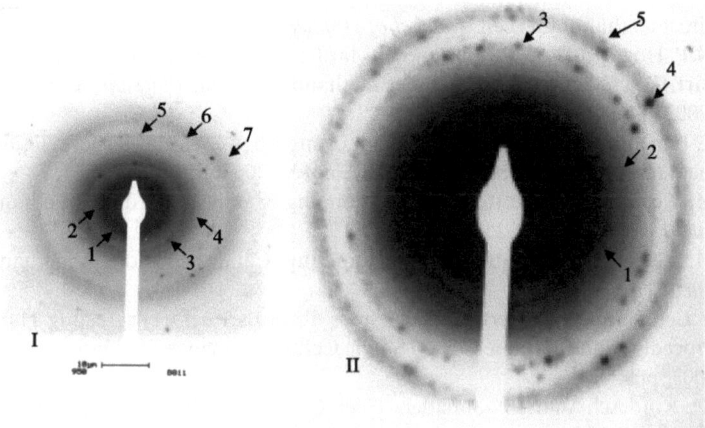

Figure 9-9. Selected area electron diffraction pattern of Ag cermet electrode

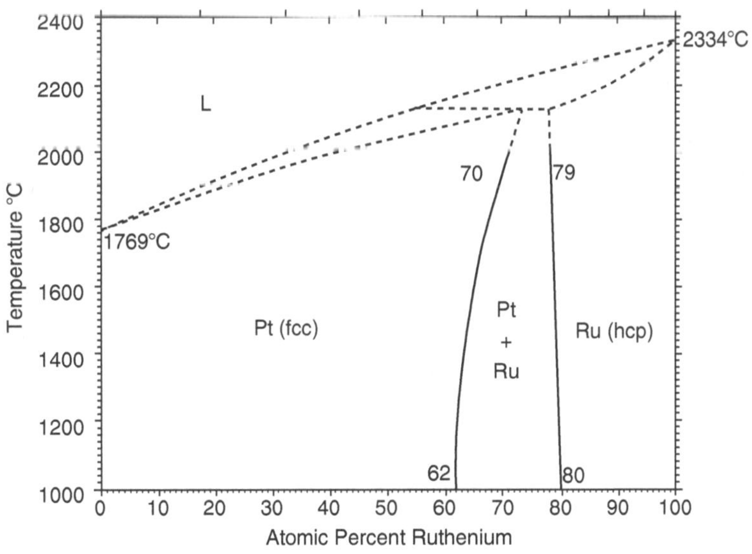

Figure 9-10. Platinum–ruthenium phase diagram

CONCLUSIONS

1. A DMFC anode was demonstrated using Ag resinate decomposed on a porous ceramic substrate as the electrode and platinum/ruthenium resinate decomposed as a catalyst structure.
2. As the processing is compatible with LTCC processing, development of fuel cells cofired into LTCC substrates are feasible.

References

[1] K. Scott, W.M. Taama, and P. Argyropoulos, Materials aspects of the liquid feed direct methanol fuel cell, *Journal of Applied Electrochemistry*, 28 (1998) 1389–1397.
[2] M.P. Hogarth, T.R. Ralph, and J. Matthey, Catalysis for low temperature fuel cells; part III: challenges for the direct methanol fuel cell, *Platinum Metals Review*, 46(4) (2002) 146–164.
[3] C. Lim and C.Y. Wang, Development of high-power electrodes for a liquid-feed direct methanol fuel cell, *Journal of Power Sources*, 2003, pp. 145–150.
[4] M. Neerat, D. Leveratto, and U. Stimming, Catalysts for direct methanol fuel cells, *Fuel Cells*, (1) (2002) 25–26.
[5] M. Neerat, D. Leveratto, and U. Stimming, Catalysts for direct methanol fuel cells, *Fuel Cells*, 2(1) (2002) 25–30.
[6] P. Lunghi and S. Ubertini, First Steps Towards Fuel Cells Testing Harmonization: Procedures and Parameters for Single Cell Performance Evaluation, Vol. 3, Suppl. 4, 2003, pp. 208–219.
[7] K. Kordesch and G. Simader, Fuel Cells and Their Applications, Wiley-VCH, Weinheim, 1996.
[8] A.S. Arico, S. Srinivasan, and V. Antonucci, DMFCs: from fundamental aspects to technology development, *Fuel Cells* 1(1) (2001) 133–161.
[9] R. Parsons and T. Van der Noot, *Journal of Electroanalytical Chemistry*, 257 (1988) 9.
[10] X. Zhang, F. Zhang, K.-Y. Chan, Synthesis of Pt–Ru–Mo ternary metal nanoparticles by microemulsions, their characterization and electrocatalytic properties, *Journal of Materials Science*, 39 (2004) 5845–5848.
[11] R.F. Service, *Science*, 271 (1996) 920.
[12] R.W. Siegel, *Physics Today*, 46 (1993) 64.
[13] F. Vigier, C. Coutanceau, A. Perrard, E.M. Belgsir, and C. Lamy, Development of anode catalysts for a direct ethanol fuel cell, *Journal of Applied Electrochemistry*, 34 (2003) 439–446.
[14] Y. Morimoto and E.B. Yeager, *Journal of Electroanalytical Chemistry*, 441 (1998) 77.
[15] A.C. Boucher, Ph.D. thesis, University of Poitiers, 2002.
[16] Y. Morimoto and E.B. Yeager, *Journal of Electroanalytical Chemistry*, 444 (1998) 95.
[17] C. Coutanceau, A.F. Rakotondrainibe, A. Lima, E. Garnier, S. Pronier, J-M. Leger, and C. Lamy, Preparation of Pt–Ru bimetallic anodes by galvanostatic pulse electrodeposition: characterization and application to the direct methanol fuel cell, *Journal of Applied Electrochemistry*, 34 (2004) 61–66.
[18] S. Motoo, M. Watanabe, and N. Furuya, *Journal of Electroanalytical Chemistry*, 160 (1984) 351.
[19] M.S. Wilson and S. Gottesfeld, *Journal of Applied Electrochemistry*, 22 (1992) 1.

[20] H. Hoster, T. Iwasita, H. Baumgartner, and W. Vielstich, Pt. Ru model catalysts for anodic methanol oxidation: Influence of structure and composition on the reactivity, *Physics Chemistry Chemical Physics*, 3 (2001) 337–346.
[21] T. Iwasita, H. Hoster, A. John-Anacker, W.F. Lin, and W. Vielstich, *Langmuir*, 16(2) (2000) 522.
[22] W. Chrzanowski, H. Kim, and A. Wieckowski, *Catalyst Letters*, 50 (1998) 69.
[23] K.A. Friedrich, K.-P. Geyzers, U. Stimming, J. Stumper, and R. Vogel, *First International Symposium on Proton Conducting Membrane Fuel Cells*, eds. A.R. Landgrebe, S. Gottesfeld, and G. Halpert, The Electrochemical Society Proceedings Series, 1995, pp. 299.
[24] C.D. Stanners, D. Gardin, and G.A. Somorjai, *Journal of Electrochemical Society*, 141 (1994) 3279.
[25] R.I. Masel, Principles of Adsorption and Reaction on Solid Surfaces, Wiley, New York, 1996, pp. 456.
[26] Y. Takasu, W. Sugimoto, and Y. Murakami, Electrocatalytic oxidation of methanol and related chemical species on ultra fine Pt and Pt–Ru particles supported on carbon, *Catalysis Surveys from Asia*, 7(1) (2003) 21–29.
[27] J.O'M. Bockris and H. Wroblowa, *Journal of Electroanalytical Chemistry*, 7 (1964) 428.
[28] M. Watanabe and M. Motoo, *Journal of Electroanalytical Chemistry*, 60 (1975) 267.
[29] B.D. McNicol and R.T. Short, *Journal of Electroanalytical Chemistry*, 81 (1977) 249.
[30] J.B. Goodnough, A. Hamnett, B.J. Kennedy, R. Manohara, and S.A. Weeks, *Journal of Electroanalytical Chemistry*, 240 (1988) 133.
[31] A. Hamnett, S.A. Weeks, B.J. Kennedy, G. Troughton, and P.A. Christen, *Berichte der Bunsen-Gesellschaft Physical Chemistry*, 94 (1990) 1014.
[32] Y. Takasu, T. Fujiwara, Y. Murakami, K. Sasaki, M. Oguri, T. Asaki, and W. Sugimoto, *Journal of Electrochemical Society,* 147 (2000) 4421.
[33] S. Guerin, B.E. Hayden, C.E. Lee, C. Mormiche, J.R. Owen, A.E. Russell, B. Theobald, D. Thompset, Combinatorial Electrochemical Screening of Fuel Cell Electrocatalysis, *J. Comb. Chem.* 6 (1) (2004) 149–158.
[34] W.H. Lizcano-Valbuena, A. de Souza, V.A. Paganin, C.A. Letiete, F. Galembeck, and E.R. Gonzalez, Performance of a DMFC using Pt–Ru/C catalysts prepared by reduction of formic acid, *Fuel Cells*, 2(3–4) (2002) 159–165.
[35] Y.-M. Tsou, L. Cao, E. De Castro, Novel High Performance PtRu Alloy Catalysts for Methanol Oxidation and CO Tolerance Applications, Abs. 675, 205th Meeting, © 2004 The Electrochemical Society, Inc., 2004.
[36] K.A. Friedrich, K.P. Geyzers, A.J. Dickinson, and U. Stimming, *Journal Electroanalytical Chemistry*, 524–525 (2002) 261.
[37] C. Coutanceau, F. Hahn, P. Waszczuk, A. Wieckowski, C. Lamy, and J.-M. Leger, Radioactive labeling study and FTIR measurements of methanol adsorption and oxidation on fuel cell, *Fuel Cells*, 2(3) (2002) 153–158.
[38] A. Wieckowski and J. Sobkowski, *Journal of Electroanalytical Chemistry*, 73 (1976) 317.
[39] M.I.S. Lopes, B. Beden, F. Hahn, J.-M. Leger, and C. Lamy, *Journal of Electroanalytical Chemistry,* 313 (1991) 323.
[40] S. Gottesfeld and T. ZAwodzinski, Ploymer Electrolyte Fuel Cells, eds. R.C. Alkire, H. Gerisher, D.M. Kolb, and C.W. Tobias, Advances in Electrochemical Science and Engineering, Wiley-VCH, Weinheim, 1997, Vol. 5, p. 195.
[41] B. Rohland and V. Plzak, *Journal of Power Sources*, 84 (1999) 183.
[42] A. Kabbabi, R. Faure, R. Durand, B. Beden, F. Hahn, J.-M. Leger, and C.Lamy, *Journal Electrochemical*, 444 (1998) 41.

[43] U.A. Paulus, U. Endruschat, G.F. Feldmeyer, T.J. Schmidt, H. Bonnemann, and R.J. Behm, *Journal of Catalysis*, 195 (2000) 383.

[44] K. Scott, W.M. Taama, and P. Argyropoulos, Materials aspects of the liquid feed direct methanol fuel cell, *Journal of Applied Electrochemistry*, 28 (1998) 1389–1397.

[45] T.E. Mallouk and E.S. Smotkin, *Combinatorial Catalyst Development Methods*, 2(3) (2003) 334–347.

[46] G. D'Arrigo, C. Spinella, G. Arena, and S. Lorenti, Fabrication of miniaturized Si-based electro catalytic membranes, *Materials Science and Engineering*, 23 (2003) 13–18.

[47] S.C. Kelley, et al., *Electrochemical and Solid-State Letters*, 3(9) (2000) 407–409.

[48] G.Q. Lu and C.Y. Wang, Development of micro direct methanol fuel cells for high power applications, *Journal of Power Sources*, 144 (2005) 141–145.

[49] Engelhard Corporation 101 Wood Avenue Iselin, NJ 08830-0770 Web site: http://www.engelhard.com

[50] X. Ren, M. Wilson, and S. Gottesfeld, *Journal of Electrochemical Society*, 143 (1996) L12.

[51] Y.-M. Tsou, L. Cao, E. De Castro, *Novel High Performance PtRu Alloy Catalysts for Methanol Oxidation and CO Tolerance Applications*, Abs. 675, 205th Meeting, © 2004 The Electrochemical Society, Inc., 2004.

[52] Y. Takasu, W. Sugimoto, and Y. Murakami, Electrocatalytic oxidation of methanol and related chemical species on ultra fine Pt and Pt–Ru particles supported on carbon, *Catalysis Surveys from Asia*, 7(1) (2003) 21–29.

Chapter 10

AUTOMATED FLUID DISPENSING FOR FUEL CELL MANUFACTURE AND ASSEMBLY

Alan Lewis
Director of Application Engineering, Asymtek, 2762 Loker Ave West, Carlsbad, CA 92010, USA

INTRODUCTION

This chapter introduces the reader to the automated dispensing equipment commonly used in electronics packaging, printed circuit board assembly, and other electronics manufacturing. Automated dispensers excel in the areas of manufacturing that can be most helpful in the construction of fuel stacks.

To demonstrate these capabilities, the discussion will describe a variety of mechanical, electrical, chemical, and thermal processes which could apply to the particular needs of fuel cell manufacturing. Typical automated dispensing processes include bonding, sealing, making electrical connections, depositing getters and catalysts, and applying insulating coatings.

Manufacturing engineers who design production lines that incorporate any of these processes must consider these parameters which influence the choice of dispensing equipment:

- Fluid characteristics
- Dispense surface geometry
- Fluid deposit geometry
- Tolerances

Before developing an automated process for dispensing fluids, it is critical to know and understand the desired results and this begins by understanding the fluid characteristics. Often it is best to collaborate early with both the fluid supplier and the supplier of the dispensing equipment.

DEFINING THE DESIRED RESULTS

When developing an automated dispensing process, the equipment suppliers need to know the desired end result. This may seem like a trivial observation, but shaping the desired result is a significant first step toward long-term success in manufacturing.

The most difficult part of this task is to determine the necessary tolerances. All dispensing processes will contain inherent variability. The tolerances will have a strong effect on the sophistication of equipment chosen, the speed of the process, and overall manufacturing cost.

Core process requirements and tolerances depend on the requirements of a specific application, but typically include:

- *Total amount of material*, in volume or weight, and allowable tolerance of the amount
- *Dot diameter, height, profile, and allowable tolerances*
- *Bead height, width, cross-section area, mass, or volume per unit length*: When dispensing a bead, consider not only the allowable variation from part to part, but also the allowable variation from line to line. This is particularly true of the start and stop points of the line, as these are often the most difficult to control.
- *Position accuracy*: the allowable tolerance on the placement of the material has a significant effect on the sophistication of the equipment and how the dispense part is handled.
- *Coating thickness*: calculate both the desired *average coating thickness* and the allowable *variation in thickness* across the coating surface. This is of particular importance near the edge of a coating if only some areas of the part are coated.
- *Keep out zones*: it is very important to define where the material must not be. Overspray for coatings and wicking of materials in noncoating applications can create problems for some applications.
- *Production speed*: this can be defined as tact time (the time to dispense on one part) or the number of units that need to be produced in a period of time.
- *Bond line thickness and flow out shape*: often fluids are compressed between two surfaces to create a mechanical bond, electrical interface, and/or thermal interface. When this is the case, the shape of the fluid as dispensed can significantly affect the bond line thickness, flow out, and coverage of the interface. Any part geometries that cause variation in the bond line thickness will likewise influence the volumetric tolerances of the material to be dispensed.
- *Metrology methods*: for all of these requirements, one should also consider the means of verifying these requirements and tolerances. Metrology considerations are beyond the scope of this chapter, but dispensing equipment suppliers are often a good source of advice for metrology equipment for specific requirements.

GENERAL FLUID CONSIDERATIONS

Specialty fluids are often engineered for their end use, but careful consideration must be given to their properties as dispensed. These include, but are not limited to:

- *Viscosity* and the factors that can change the viscosity
- *Specific gravity* or density of the fluid
- *Chemical compatibility with wetted parts* of dispensing equipment
- *Mechanical compatibility with wetted parts* of the dispensing equipment such as abrasion or sealing issues
- *Suspended particles or solids* (shape, hardness, fill ratios, tendency to settle)
- *Volatile components* that can evaporate or dissolve with time
- *Pot life* of the material
- *Curing mechanism* (heat, chemical, evaporation, UV)
- *Storage and handling* of the fluid
- *Health and safety considerations*, such as toxicity and flammability
- *Compatibility with the manufacturing environment*, such as sensitivity to light, moisture, oxygen, or temperature
- *Manufacturing variations in the fluid* that can affect any of the above characteristics from batch to batch

TYPES OF DISPENSING TECHNOLOGY

The automated equipment discussion begins with the valve and pump technologies used for depositing fluids. These technologies can be broadly categorized as needle dispensing, coating technologies, and jetting.

Needle Dispensing

Needle dispensing is described as extruding material under pressure to deposit dots or lines of materials and can be used for a wide variety of materials. In all cases, the tip of the needle or nozzle is positioned close to the dispensing surface and a combination of gravity and surface tension pull the material away from the dispensing tip.

Discrete deposits or dots can be created by holding the dispense tip stationary relative to the dispensing surface. Lines or beads of material can be created by moving the tip relative to the surface while extruding the material. The difference between needle dispensing technologies lies in the mechanism used to create the extrusion pressure.

Characteristics of Time-Pressure Dispensing

- ***Flow rate***
  - Proportional to viscosity & pressure
  - Inversely proportional to needle length
  - Proportional to needle ID^4
- ***Simple operation at the expense of accuracy***

$$Q = \frac{\pi d_n^{\,4}}{128 \mu_n \, l_n} (P_2 - P_3)$$

Q = Total flow rate
μ_n = Material viscosity in needle
P_2 = Pressure in needle hub
P_3 = Pressure at needle outlet
l_n = Length of needle
d_n = Inside diameter of needle

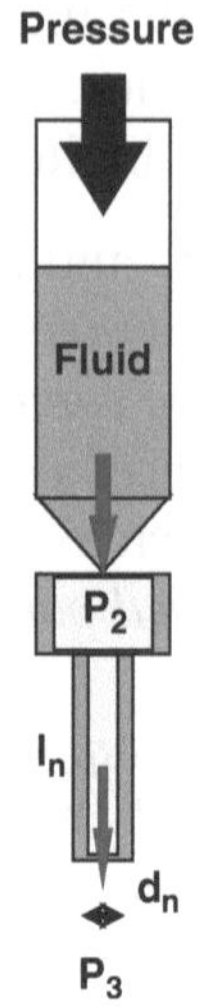

Figure 10-1. Time–pressure dispensing

One of the simpler methods is time–pressure dispensing, shown in Figure 10-1.

In time–pressure dispensing, the fluid is pressurized pneumatically. The amount of fluid dispensed is controlled by cycling pressure during the dispense cycle.

Alternatively, the pneumatic pressure can be held constant and a valve used to control the fluid flow. Needle valves, spool valves, and rotor valves are commonly used in this type of device. In all devices of this type, the fluid flow rate (Q) is directly proportional to the material viscosity. Those factors that change the viscosity should be controlled carefully or the equipment must be able to compensate for the changes in order to achieve consistent results.

Materials that are sensitive to moisture or oxygen can be pressurized with dry nitrogen. Some materials, such as many silicones, can absorb gas in solution under pressure. This can create problems, such as making the fluid compressible and difficult to control, and may sometimes lead to bubbles in the dispensed fluid. If that happens, it is best to use another method of extruding the material.

Two advantages of time–pressure dispensing are:

1. It can be very gentle on sensitive or fragile fluids that are affected by shear stress and strain.
2. It is a simple technology compared with some other dispensing methods.

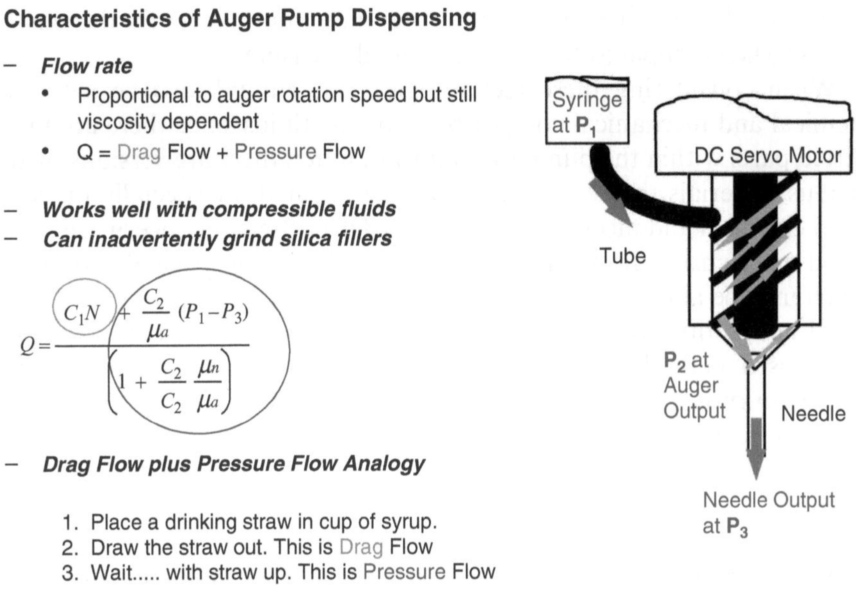

Figure 10-2. Auger pump dispensing

The primary disadvantage is that time–pressure dispensing is less precise than some other dispensing methods.

Auger pumps and *piston pumps* are also commonly used in needle dispensing. Examples are shown in Figures 10-2 and 10-3.

While more complex and expensive to purchase and maintain, these pumps are commonly used to improve processing speed and accuracy. They are generally much less sensitive to viscosity changes in material, can

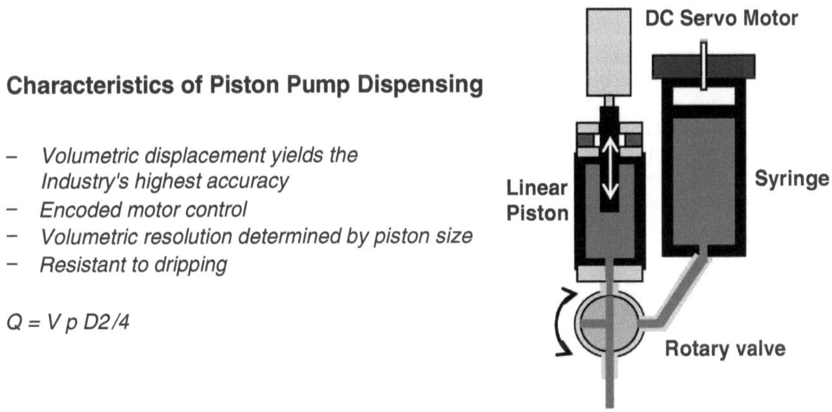

Figure 10-3. Piston pump dispensing

be more easily calibrated, and give the user more control over the dispensing process when compared to time–pressure dispensing.

When considering these technologies, one should also be aware of chemical and mechanical compatibility of the fluids since there are more wetted parts within the pump mechanisms. Often there are versions of the various materials that have been hardened, coated, or specially made to avoid chemical and mechanical problems. Of particular concern are fluids that use UV curing mechanisms, aggressive solvents, or suspended particles which may be abrasive.

Although time–pressure, auger, and piston systems are the most commonly used methods of needle dispensing, some other pump designs exist. For the purposes of fuel cell stack assembly, however, the considerations listed above apply to most technologies.

Jetting

Jetting technology is one of the more recent technologies in automated fluid dispensing and is quickly evolving. These technologies eject discrete volumes of materials from a nozzle and do not rely on the surface tension of the fluid on the dispensed surface to pull the fluid away from the nozzle.

The leading dispense jetting technology available today utilizes a pneumatic piston with a ball tip to impel fluid through a narrow orifice at the end of the jet nozzle as shown in Figure 10-4.

Air pressure raises the piston, allowing fluid to flow around the piston into the nozzle. When air pressure is removed, a spring returns the piston so that the ball again seats in the nozzle orifice.

Characteristics

- ***Discrete volumetric accuracy***
- ***Smallest flow out area of all dispensing methods***
- ***Able to dispense into narrow cavities***
- ***No dripping***
- ***Shoot from above the die***
- ***Often faster throughput***

*Q = dots/sec * mg/Dot*

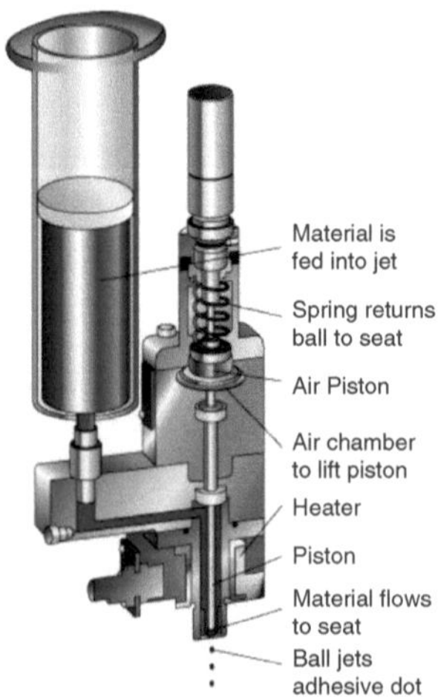

Figure 10-4. Jet dispensing

As the ball reseats, it shoots a droplet of fluid out of the end of the nozzle. Adjusting the nozzle orifice, air pressure and fluid pressure controls the size of droplets. Precise heat control at the nozzle tip maintains fluid temperature at an optimum viscosity for jetting and reduces variation in production.

The jetting mechanism propels small dots at rates up to 200 Hz in four software-controlled modes: *distance based*, *time based*, *fixed number of dots per line*, and *continuous line with breaks*, enabling the construction of many different sizes and shapes of dots and lines. Because the momentum of the fluid comes from the jetting action, proximity to the substrate (dispense gap) is much less critical than it is in needle dispensing.

The inherent simplicity of the jetting mechanism enables it to adapt to a wider variety of adhesive fluids, dispense patterns, and circuit board geographies than is possible with a needle dispenser. This simplicity facilitates multiple processes using a single jetting dispenser.

For most users, the most attractive feature of jetting is the increased speed compared with needle dispensing. The primary reason for jetting to be faster than needle dispensing is the reduced amount of mechanical travel required by the jet mechanism.

Whereas a needle dispenser often requires a stand-off to provide dispense gap feedback to the operating system, jetting does not require the dispense tip to contact the surface of the part on which fluid is dispensed, which gives jetting another significant advantage over needle dispensing. In some cases, this advantage can be the difference that enables a technology to use dispensing. One example is the ability to jet materials in small spaces where needles will not fit.

Coating

For many applications, a thin coating of material over a relatively large area is required. The most common types of coating are film coating (nonatomized) and atomized spraying.

Film coating equipment creates a curtain of nonatomized fluid for covering a surface. Film coating can be applied at a relatively high speed (up to 0.7 m s^{-1}) and has the advantage of being very selective with virtually no overspray. Typical coating widths for a single pass are approximately 20 mm. The coating width and material flow rate can be calibrated and controlled with some equipment. The disadvantage of this method is that the wet coating thickness is usually greater than 75 μm and the viscosity of the material is limited to approximately 150 cps (Figure 10-5).

Air atomization, airless atomization, and ultrasonic atomization are some of the methods used to create atomized spray. While these methods differ slightly, they all are generally less selective than film coating and the

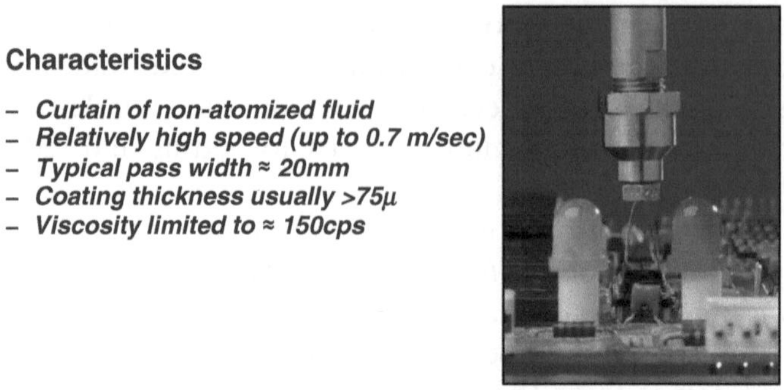

Figure 10-5. Film coating

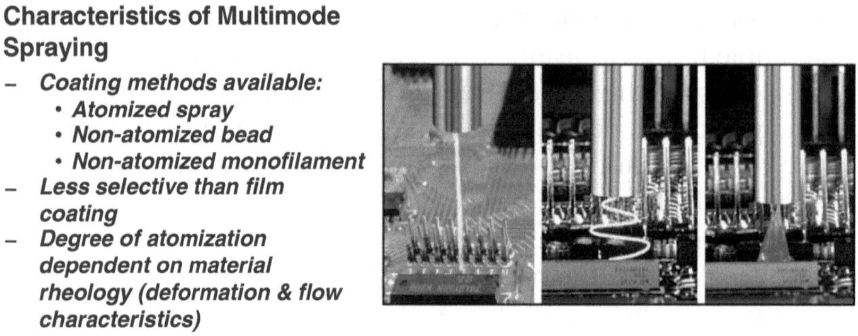

Figure 10-6. Multimode spraying

degree of atomization is highly dependent on the material rheology. Care must be taken to avoid problems with overspray and ventilation.

To offer more flexibility in coating applications, one equipment supplier has developed a coating applicator capable of a wide range of viscosities (30–3,500 cps) and three dispense options: atomized spray, nonatomized bead, or nonatomized monofilament (Figure 10-6).

AUTOMATED EQUIPMENT FEATURES

Robotic Motion

Most of the equipment used in automated dispensing in the electronics industry is designed to place fluid on flat or nearly flat surfaces. This significantly simplifies the process since the motion of the robot during the dispensing operation can be limited to two axes. Robots that can dispense

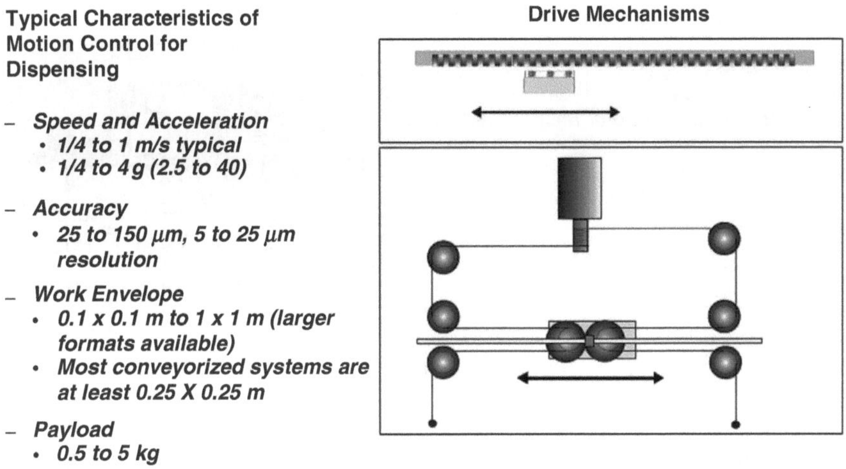

Figure 10-7. Motion control for dispensing

on 3D surfaces are readily available, but this can affect the cost, complexity, and speed of the dispensing process (Figure 10-7).

There are a variety of drive mechanisms commonly used automated dispensing including cable drive, lead screw, and linear motors. Most modern equipment uses closed loop servo mechanisms, although open loop stepper motors can be found in some less expensive equipment. The drive mechanism is not as important as the overall speed and accuracy of the equipment. Other factors, such as the alignment process and dispensing technology often overshadow the characteristics of the drive mechanism.

Part Handling and Alignment

Automated Part Handling

Designing a dispense process requires a decision to load parts into the dispenser manually or to use an automated loader. If the parts are to be loaded manually, tooling can be configured with the equipment supplier. If the parts are to be loaded automatically, time and money can be saved by following one of the equipment mechanical and electrical interface standards.

The most common equipment interface standards include those of SEMI (http://www.semi.org/), SMEMA (http://www.smema.org/smema-standards.htm), or IPC (http://www.ipc.org/). Often, parts can be configured to be loaded into a tray or "boat" which facilitates handling, loading, and unloading into standard cassettes that can be transferred to other

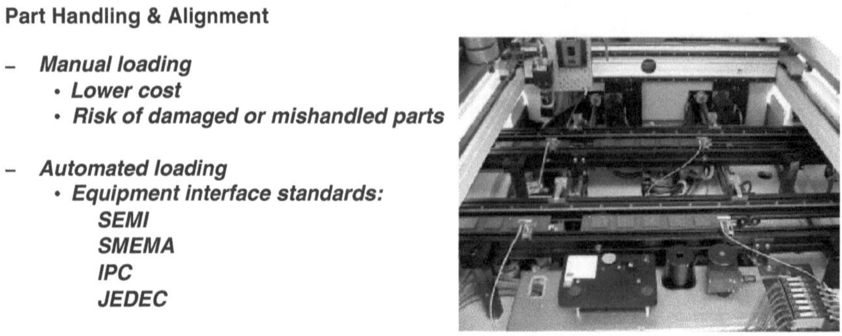

Figure 10-8. Handling parts

assembly equipment such as plasma cleaners, pick and place equipment, and ovens.

JEDEC standard trays and boats manufactured by Auer Precision (http://www.auerprecision.com/index.html) are common in the industry (Figure 10-8).

For larger stack panels, one might investigate the emerging standards being developed in the Flat Panel Display industry. Professional organizations in the flat panel industry such as SPIE (http://www.spie.org/) and SID (http://www.sid.org/) are good sources of information.

Since the Flat Panel Display industry must confront the problems of bonding and sealing relatively large areas with special atmospheric requirements (vacuums and/or special process gases), the equipment suppliers in this industry also may serve as a good source of information.

Temperature Control

It is often necessary to control the temperature of a part during automated fluid dispensing in order to gain control of the process. Heating the parts enhances material flow for encapsulation or underfill processes, for example, which are described later in this chapter.

If an application requires part heating, dispensing equipment suppliers offer infrared (IR) heaters, heated contact tooling, or convection heaters. IR heaters can produce a fast temperature rise on a tightly focused area; heated contact tooling is gentler heating, but has a slower ramp-up time than IR; convection heaters use heated gas to evenly raise the temperature of multiple parts with dissimilar mass.

Most dispensing equipment suppliers offer standard equipment capable of heating parts from ambient to approximately 140°C. Higher temperatures may require customized heating and insulating options.

Often an application also requires controlling the fluid temperature during dispense. This can be caused by the need to:

- Cool the fluid to extend working life in production
- Raise or lower the viscosity of the fluid to
  - Increase the speed of delivery
  - Shape the temperature profile of the dispense fluid
- Induce phase change in a material

Automated Vision

In order to place fluid accurately and consistently within the desired tolerance, it is often necessary to use automated vision systems to align the parts to the dispensing robot. This can add flexibility to a system and reduce the complexity of "hard tooling" which requires pins or mechanical alignment devices.

These vision systems use a variety of pattern recognition routines to find alignment marks or features on the part to place the fluid where it is desired. Incorporating edge, color, or other distinguishing features into the parts can facilitate this process (Figure 10-9).

Automated Fluid Calibration

For critical applications where the fluid volume must be precisely controlled over long periods of time, automated fluid calibration may be required. Both fully or semiautomated systems use volume measurement

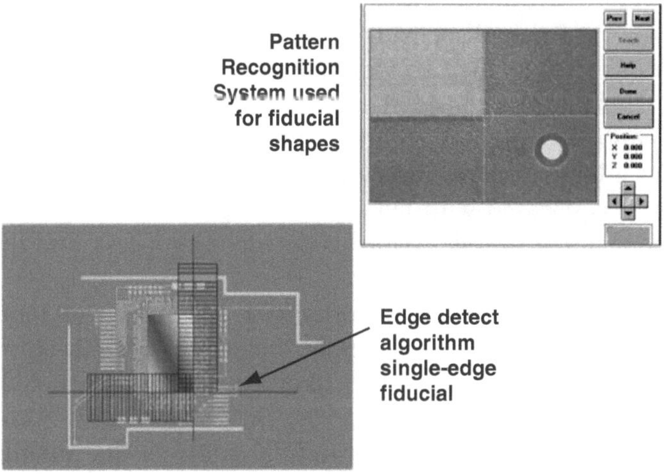

Figure 10-9. Automated vision

or mass measurement to dispense in the nanoliter and microliter range. Coating and spraying systems typically use closed loop flow control to manage fluid volume.

FLUIDS USED IN AUTOMATED DISPENSING

Adhesives and Encapsulants

Adhesives and encapsulants are dispensed as fluids and are then polymerized (or cured) to a solid state in their end use form. This is a broad range of materials, usually categorized by the curing mechanism as well as the basic chemistry.

Adhesives are designed to bond parts, whereas encapsulants are designed to protect components from mechanical, thermal, and/or chemical influences.

The following characteristics determine the material of choice for an adhesive or encapsulant application:

- Adhesion strength
- Coefficient of thermal expansion (CTE)
- Modulus of elasticity
- Chemical resistance
- Cure speed

Epoxies are popular because of their strength and cure speed. *Two part epoxies* often complicate the dispensing process, but simplify the curing process. They are frequently lower cost, but must be mixed in small batches or mixed "on the fly" in dispensing equipment. Because the curing reaction commences as soon as the two materials contact, the viscosity changes as they cure. This must be considered in the manufacturing process.

Single part epoxy presents another option. Some materials, such as surface mount adhesives, have the curing catalyst locked in a phase change material that is released with heat. Cure temperatures range from 100 to 150°C. The material supplier controls the mixing process, and these materials have excellent shelf life.

Some single part epoxies are premixed and frozen at temperatures ranging from –5 to –40°C. Freezing the material essentially stops the curing reaction. They are designed to have a working life of 4–24 h after thawing to ambient temperature.

Because the material supplier carefully controls the mixing process, these epoxies have very consistent properties, but handling and storage require special attention. Suppliers provide these fluids with various cure times and temperatures, but they are usually in the 90–150°C temperature range and 5 min to several hours cure time. Fluids with lower cure temperatures require longer cure times while those cured at higher temperatures require less time.

UV cure materials fit best in situations that call for low outgassing or fast curing. Special cure ovens or light wands produce the proper intensity and wavelength of light necessary to initiate the curing process. In some applications where only part of the bond area can be exposed to UV light, a secondary heat or moisture cure mechanism can be used.

Silicone-based chemistries match well with applications that require a low modulus of elasticity or high-temperature resistance. They also correspond with applications that require a combination of bonding and sealing. Curing mechanisms can include heat, moisture, or solvent evaporation. Room temperature vulcanization (RTV) materials are designed to polymerize as a solvent evaporates.

Electrically Conductive Materials

Electrically conductive materials can be made to be either isotropically conductive (IC) or anisotropically conductive (AIC). IC materials are the most common and conduct electricity in all directions. AIC materials conduct materials in one axis when compressed between two contact points.

Most IC materials are either solder pastes or silver epoxies. Besides providing an electrical path for a circuit, they can also be used to attach radio frequency shields.

Solder paste consists of solder powder suspended in a flux. The material is dispensed as a slurry, then heated to its reflow temperature, and finally cooled to create a solid solder joint. The liquid flux serves multiple purposes:

- It is a carrier for the solder during dispensing
- It is a reducing agent during reflow to prevent oxidization
- It provides good wetting for good electrical connection

These materials are predominantly used in printed circuit board assembly, and the equipment for their dispensing and reflow is well standardized. Solder paste joints provide excellent electrical and thermal conductivity, and create a strong mechanical bond.

Solder paste is specified by:

- Powder particle size, which is standardized as "Type"
- Metal alloy, which determines the reflow temperature
- Flux chemistry
- Percentage of metal content by weight of the paste

Silver epoxies consist of silver flakes suspended in an adhesive carrier. The materials are designed so that the silver flakes touch each other when cured and provide an electrically conductive path. Generally, silver epoxies have lower conductivity than solder paste, but usually have the advantage of a reflow temperature lower than the temperature required for solder paste.

They can also be designed with lower modulus of elasticity and various coefficients of thermal expansion to suit applications. Silver epoxies also feature good thermal conductivity.

AIC materials contain a relatively low percentage of conductive particles suspended in an adhesive. They make electrical contact when compressed between two electrical surfaces and conduct electricity in one axis. The particles must be size matched to the bond line thickness or they will create shorts between lines.

The adhesive contracts during cure to maintain electrical contact. These materials usually have lower conductivity than IC materials and are used to make contact between a grid of contacts that must have isolated electrical paths.

Thermal Interface Materials

Thermal interface materials are slurries of thermally conductive particles, usually diamond or metallic oxides, suspended in liquid or low modulus adhesive. They are designed to be flexible or slightly fluid to maintain good thermal contact during temperature cycling.

Because the thermal conductivity is directly related to the solid content, they are very viscous and can be abrasive to mechanical components of the dispensing equipment. They are most often used to improve the performance of heat sinks. They can be particularly useful in fuel cell applications to remove heat from reaction sites.

Conformal Coating Material

Conformal coatings protect electronic circuitry from moisture, dust, chemicals, solvents, and other harsh environments. Although usually only a few mils thick, they also dampen the effects of mechanical and thermal stresses, vibrations, and electrical noise. Conformal coatings also have been used to provide electrical insulation by inhibiting dendrite growth, a potential cause of short circuits.

Conformal coatings can be found in automotive, consumer appliance, and military applications where they improve reliability. They are designed to be applied in film thicknesses typically ranging from 20 to 150 μm. The polymerization mechanism can be UV radiation, solvent evaporation, moisture, or heat. In fuel cell applications, they can allow the designer to place the electronics in closer proximity to the fuel cell stacks.

Underfill Materials

Underfill materials are a special class of encapsulants that provide a mechanical interface to flip chips and BGA components. Although they are designed for a specific application in circuit board and processor packages, they have found other uses. The primary characteristic of these fluids that sets them apart from other encapsulants is that they are designed to flow into gaps of 10–75 μm, allowing a designer to provide a very strong bond between closely spaced gaps.

Getters and Catalyst Fluids

Getter and catalyst materials are often suspended in a fluid to form a slurry to facilitate their application. Compared to the materials listed above, there are relatively few off the shelf products and are often designed for specific applications.

They are very often found in MEMS devices when the use of automated dispensing equipment makes it easier to place precise amounts of materials in areas where it would be difficult or impossible to apply them by other methods. By applying them in liquid form with automated dispensing equipment, the amount and placement of the material can easily be adjusted to suit the needs of the task at hand.

EXAMPLES OF TYPICAL AUTOMATED DISPENSING APPLICATIONS

Encapsulation

Purpose

Encapsulation protects components from the impact of mechanical, thermal, and/or chemical stresses by enveloping the components in one of many available materials (Figure 10-10).

Characteristics of Dam & Fill Encapsulation

- **Dams usually dispensed with auger pump. Good bead width control.**
- **Dams usually need to be +/– 30 μm in height.**

- **Fill usually dispensed with piston pump. High speed and good accuracy.**
- **Fluid volume usually needs to be with +/– 2%.**

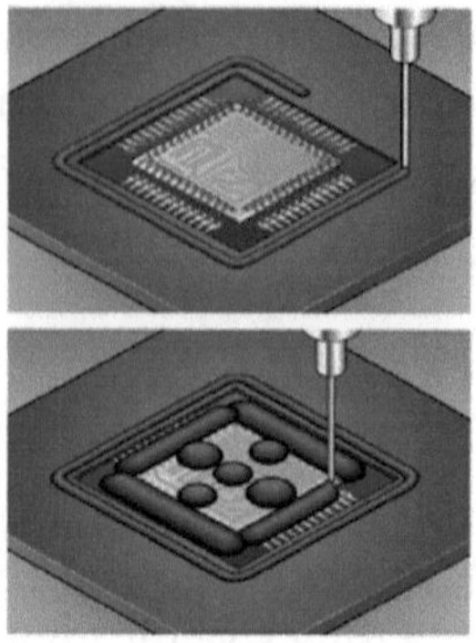

Figure 10-10. Encapsulation

Encapsulation Materials

Characteristics of most encapsulants include:

- CTE of 15–30 ppm per OC
- E of 4–10 Gpa
- Silica fillers 20–60% by volume
- High-surface tension/good wetting
- Shrink during cure
- High-electrical resistance/low ionic content
- Very high adhesion

Encapsulation Dispense Technology

1. Auger pumps are good for dam dispense because of bead control capability
2. Piston pumps
   - Good for fill because of higher speed and better accuracy than auger pumps
   - Good resistance to abrasion compared with auger pumps
3. Part preheating to 80–110°C is often assists the dispense process
4. Needle heating
   - Helps stabilize dam dispense
   - Reduces stringing on fill dispense
   - Has no effect on flow

Encapsulation Process Considerations

- Accuracy of fluid dispense is dictated by the package height.
- Fluid volume generally must be maintained within ±2 to ±10% of target height, depending on how the package interfaces mechanically.

Conformal Coating

Purpose

Conformal coatings protect many different types of products with a nearly impermeable barrier against moisture, dust, and harmful gases. In many applications they also act as dampers against shock and vibration.

The most advanced selective film coaters apply solvent-based materials in thicknesses of 0.5–8.0 mils, and apply 100% solids in thicknesses of 4.0–8.0 mils (Figure 10-11).

Conformal Coating Materials

Conformal coatings typically are either solvent-based or 100% solids:

- Solvent-based materials are relatively inexpensive and their viscosity can be easily controlled through dilution with volatile solvents.
- 100% solid materials, though more expensive, offer advantages regarding worker health, environmental restrictions on VOC emissions, and hazardous waste disposal. Because 100% solids-based coatings are actually liquids, a more accurate term might be "solvent-less."

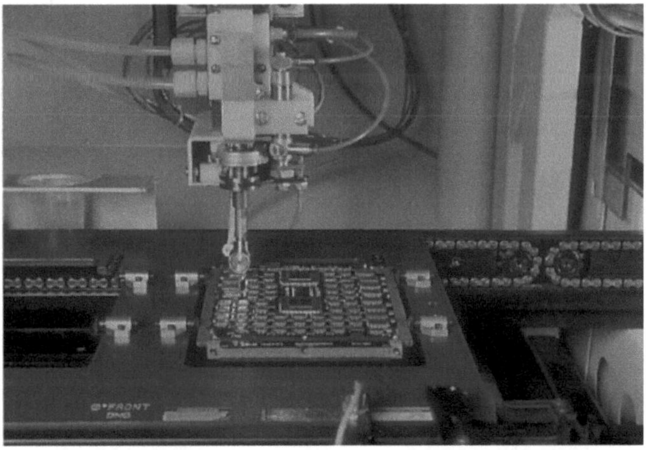

Figure 10-11. Conformal coating

Conformal coatings are usually made from one of five basic resin chemistries:

- Acrylic (AR)
- Polyurethane (UR)
- Epoxy (ER)
- Silicone (SR)
- Polyparxylylene (XY)

Conformal Coating Process Considerations

1. Evaluate a conformal coating application with the goal of prioritizing the types of protection required
   - Typical conformal coating materials offer protection against some environmental hazards, but no single fluid offers protection against all of them.
   - Cost, ease of handling, ease of application, shelf and pot life, strength and durability of the cured coating, impermeability to moisture and dust, and other factors must be considered.
2. Collaborate with both material suppliers and equipment manufacturers during the design phase of a product that will be conformally coated.

Die Attach

Purpose

During die attach, an adhesive creates a mechanical interface between the die and substrate in a semiconductor assembly. A typical dispense pattern has intersecting lines with strategically placed dots so that after die placement, the fluid is evenly distributed (Figure 10-12). Most die attach adhesives also provide thermal and/or electrical conductivity between the die and substrate.

Figure 10-12. Die attach dispensing

Die Attach Materials

The ideal die attach material has a short cure time at a relatively low-cure temperature, exhibits low moisture absorption, provides good adhesion to multiple surfaces, contributes low stress to the die and can be processed within a very tight tolerance. Many epoxies are now available meet some of these demands to a high degree, yet each material has limitations.

Die Attach Dispense Technology

- Piston pumps work well with many die adhesives, especially those with high-silver content.
- Needle heat control can control stringing and stabilize viscosity.
- Jet dispensing can deliver advantages in speed, accuracy, and repeatability in many die attach applications.
- Tighter size and space tolerances frequently make an automated vision system necessary.
- Where volumetric accuracy and repeatability are critical, mass flow calibration can be advantageous.

Thermal Interface

Purpose

Thermal interface compounds conduct heat from a part to a heat sink or lid (Figure 10-13). In fuel cell assemblies they could be used to remove heat from reaction sites.

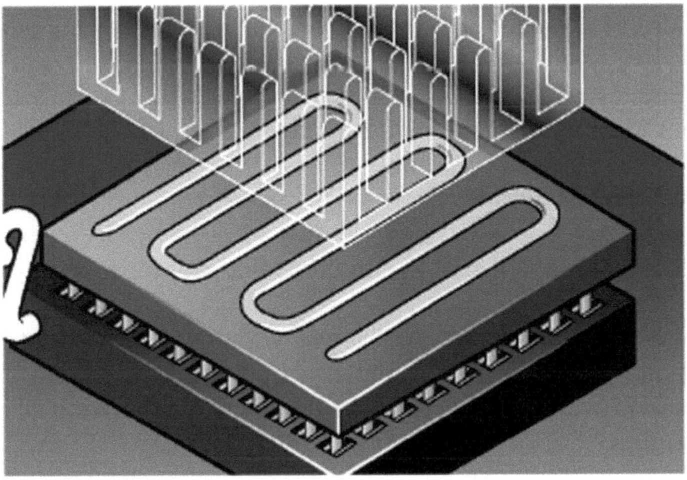

Figure 10-13. Thermal interface

Thermal Interface Materials

Thermal compounds are typically highly filled with silver, silica, alumina oxide, or diamond powder and usually have a high viscosity.

Thermal Interface Dispense Technology

- Time–pressure dispensing is often used in thermal interface applications, but lacks accuracy compared with other available technologies, with typical tolerances of ±20%
- Auger pumps offer both higher flow rates and better accuracy than time–pressure, with a typical tolerance of ±2%

Thermal Interface Process Considerations

- Care must be taken so that the thermal interface material does not conduct stress to the die.
- High viscosity combined with high-solids content of most thermal interface fluids causes abrasion of mechanical components of dispense equipment. Worn dispense components affect overall dispense quality, especially volume control.

Gaskets and Seals

Purpose

Gaskets and seals form a thin, protective intermediate layer of material between two surfaces. Often there is a significant difference between the modulus of elasticity and the CTE of the materials being gasketed or sealed.

The gasket or seal alleviates these differences in materials to protect them from warping or breaking during normal use of the final product.

Gasket and Seal Materials

- UV cure materials fit best in situations that call for low outgassing or fast curing.
- Silicone-based chemistries match well with applications that require a low modulus of elasticity or high-temperature resistance. They also correspond with applications that require a combination of bonding and sealing.

Gasket and Seal Dispense Technology

- Time–pressure dispensing can be used effectively on many simple, low value-added applications.
- Jet dispensing is becoming a preferred technology for more sophisticated applications such as flat panel displays
- Gaskets and seals are normally created by dispensing a line of fluid that is compressed between two parts (Figure 10-14, top photos).
- It is not always necessary, however, to dispense a continuous line if it is being compressed between two parts (or laminated). In fact, it may be faster and give better sealing around corners if you do noncontinuous lines (Figure 10-14, bottom photos).
- Jetting enables the use of a series of precisely placed dots that form a line when compressed (Figure 10-15).

Gasket and Seal Process Considerations

- It is important to calculate the volume per unit length of the *final compressed line.*

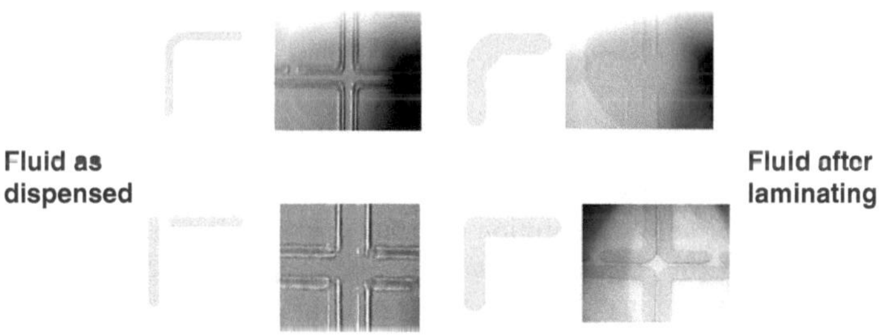

Figure 10-14. Gaskets and seals

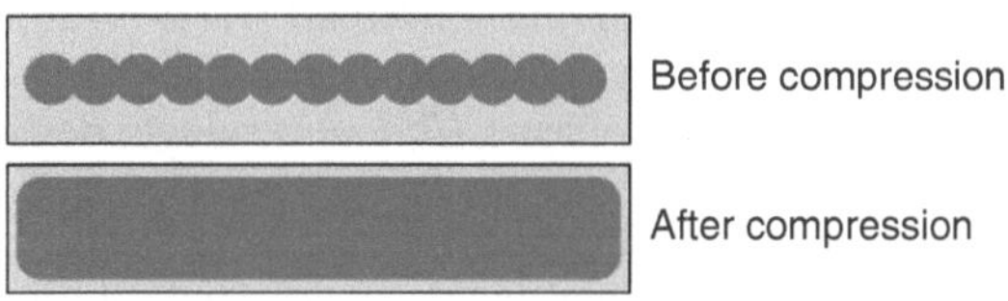

Figure 10-15. Jetted dots used in seal dispensing

SUMMARY

Automated dispensing is a mature, proven technology that serves many purposes in a wide range of industries. Dispensing equipment manufacturers regularly collaborate with fluid vendors to supply machines and materials appropriate to applications that range from simple to highly advanced technology.

Many of these companies have sophisticated applications labs with highly trained, experienced applications engineers who will work with engineers to develop a dispense process that precisely fits the needs of their customers.

Although the fuel cell industry is a rapidly evolving technology that presents novel challenges in the area of dispensing, these applications engineers are prepared to help fuel cell manufacturers develop new processes that fit their unique requirements.

ACKNOWLEDGEMENTS

The author would like to acknowledge the contribution of Alec Babiarz, Horatio Quinones, Thomas Ratledge, Floriana Suriawidjaja, and Rick Zakrajsek for their contribution to the material contained in this chapter.

References

[1] A. Lewis, U.S. patent, 1968.
[2] F. Savart, Memoire sur la constitution des veines liquides lancees par des orifices circulaires en mince paroi. *Annales de Chimie et de Physique* 53 (1833) 337–386.
[3] J.W.S. Rayleigh, On the instability of jets. *Proceedings of the London Mathematical Society*, 10(4) (1879) 4–13.
[4] J.W.S. Rayleigh, On the capillary phenomena of jets. *Proceedings of the Royal Society*, 29 (1879) 71.
[5] J.W.S. Rayleigh, On the instability of a cylinder of viscous liquid under capillary force. *Philosophical Magazine*, 34 (1892) 145–154.
[6] K.C. Chaudhary, I.G. Redekopp, and T. Maxworthy, The non-linear capillary instability of a liquid jet. *Journal of Fluid Mechanics* 96(part II) (1979) 257–312.
[7] W.T. Pimbley, Drop formation from a liquid jet: a linear one-dimensional analysis considered as a boundary value problem. *IBM Journal of Research Development,* 29 (1984) 148–156.
[8] C.W. Hansell, Jet sprayer actuated by supersonic waves. U.S. patent 2,512,743, 1950.
[9] D.B. Bogy and F.E. Talke, Experimental and theoretical study of wave propagation phenomena in drop-on-demand ink jet devices. *IBM Journal of Research Development,* 29 (1984) 314–321.

[10] J.F. Dijksman, Hydrodynamics of small tubular pumps. *Journal of Fluid Mechanics,* 139 (1984) 173–191.
[11] R.L. Adams and J. Roy. A one-dimensional numerical model of a drop-on-demand ink jet. *Journal of Applied Mechanics,* 53 (1986) 193–197.
[12] D.B. Wallace, A method of characteristics model of a drop-on-demand ink-jet device using an integral method drop formation model. ASME publication 89-WA/FE-4, 1989.
[13] J.R. Pies, D.B. Wallace, and D.J. Hayes, High density ink jet printhead. U.S. patent 5,235,352, 1993.
[14] J. Roy and J.S. Moore, Drop-on-demand ink jet print head. U.S. patent 5,087,930, 1992.
[15] P.A. Hoisington, R.R. Schaffer, and K.H. Fischbeck, Ink jet array. U.S. patent 4,835,554, 1989.
[16] M. Usui and T. Katakura, Actuator for an ink jet print head of the layered type with offset linear arrays of pressure generating chamber. U.S. patent 6,033,058, 1996.
[17] J.S. Aden, J.H. Bohorquez, D.M. Collins, M.D. Crook, A. Garcia, and U.E. Hess, The third generation hp thermal inkjet printhead. *Hewlett-Packard Journal* 45(1) (1994) 41–45.
[18] J.M. Waldvogel, D. Poulikakos, D.B. Wallace, and R.M. Marusak, Transport phenonmena in picoliter size solder droplet dispensing on a composite substrate. *Journal of Heat Transfer* 118 (1996) 148–156.
[19] R.D. Deegan, O. Bakajin, T.F. DuPont, G. Huber, S.R. Nagel, and T.A. Witten, Capillary flow as the cause of ring stains from dried liquid drops. *Nature,* 389 (1997) 827–829.
[20] 1998 National Electronics Manufacturers Initiative Roadmap, 1998. www.inemi.org
[21] N. Van Veen and D. Schwarzbach, Solderjetting, A Software Driven Technology for Maskless Waferbumping. *Proceedings, IMAPS International Symposium on Microelectronics* (Chicago), 1999, pp. 154–159.
[22] C.W. Argento, T. Flynn, and C. Demers, Next Generation Solder Jetted Wafer Bumping for Very Fine Pitch Flip Chip Technology Applications and Beyond. *Proceedings, IMAPS International Symposium on Microelectronics,* Chicago, 1999, pp. 160–165.
[23] R. Godin, S. Pearson, and R. Lasky, A Novel Process for Solder Deposition. *SMT Magazine*, 1 (1997) 66–68.
[24] E.P. Muntz, M. Orme, G. Pham-Van-Diep, and R. Godin, An Analysis of Precision, Fly-Through Solder Jet Printing for DCA Components. *Proceedings, ISHM '97*, Philadelphia, 1997, pp. 671–680.
[25] J. Priest, E. Jacobs, C. Smith, P. DuBois, B. Holt, and B. Hammerschlag, Liquid metal-jetting technology: application issues for hybrid technology. *International Journal of Microcircuits and Electronic Packaging*, 15(3) (1994) 219–227.
[26] C. Hendricks, Inertial confinement fusion targets. *Proceedings of the 2nd International Colloquium on Drops and Bubbles*. NASA-CR-168848, JPL, 1982, pp. 82–93.
[27] T.M. Smith and R.E. Winstead, Electrodynamic pump for dispensing molten solder. U.S. patent 5,377,961, 1995.
[28] T. Schiesser, E. Menard, T. Smith, and J. Akin, Microdynamic solder pump: drop on demand eutectic SnPb solder dispensing device. *Proceedings, Surface Mount International*, San Jose, CA, 1994, 501–509.
[29] D.J. Hayes, D.B. Wallace, M.T. Boldman, and R.M. Marusak, Picoliter solder droplet dispensing, *Microcircuits and Electronic Packaging,* 16(3) (1993) 173–180.

[30] D.J. Hayes, W.R. Cox, and M.E. Grove, Microjet printing of polymers and solder for electronics manufacturing, *Journal of Electronics Manufacturing*, 8(3 and 4) (1998) 209–216.
[31] D.B. Wallace, Method and apparatus for forming microdroplets of liquids at elevated temperatures. U.S. patent 5,415,679, 1995.
[32] Advanced Embedded Passives Technology final report. http://aept.ncms.org/about.htm
[33] http://www.darpa.mil/dso/trans/mice.htm
[34] E. Reddington, A. Sapienza, B. Gurau, R. Viswanathan, S. Sarangapani, E.S. Smotkin, and T.E. Mallouk, Combinatorial electrochemistry: a highly parallel, optical screening method for discovery of better Electrocatalysts, *Science*, 280(5370) (1998) 1735–1737.
[35] J.L. Fernández, V. Raghuveer, A. Manthiram, and A.J. Bard, Pd–Ti and Pd–Co–Au electrocatalysts as a replacement for platinum for oxygen reduction in proton exchange membrane fuel cells. *Journal of American Chemical Society*, 127(38) (2005) 13100–13101.
[36] http://www1.eere.energy.gov/hydrogenandfuelcells/pdfs/tiax_platinum.pdf
[37] Sanders Prototype, Inc., 2000. www.sanders-prototype.com
[38] 3D System, 2000. www.3dsystems.com
[39] Generis GmbH, www.generis.de
[40] Packard Bioscience, www.packardbioscience.com
[41] http://www.microfab.com
[42] http://www.microdrop.com
[43] http://gesim.de/home.htm
[44] http://www.dimatix.com/
[45] http://displaysearch.com
[46] K. Aschheim, Gene detection by array. *Nature Biotechnology*, 18 (2000) 1129.
[47] S.P.A. Fodor, J.L. Read, M.C. Pirrung, L. Stryer, A. Tsai Lu, and D. Solas, Light-directed, spatially addressable parallel chemical synthesis. *Science*, 251 (1991) 767–773.
[48] http://www.plasticlogic.com

Chapter 11

INK-JET AS DIRECT-WRITE TECHNOLOGY FOR FUEL CELL PACKAGING AND MANUFACTURING

Virang G. Shah, Donald J. Hayes, and David B. Wallace
MicroFab Technologies, Inc., Plano, Texas, USA

INTRODUCTION

Ink-jet printing is being widely used and seriously considered for dispensing and printing a variety of materials. It can reproducibly dispense spheres of fluid ranging from 25 to 125 μm in diameter (8 pL to 1 nL) at rates up to 25 kHz on-demand, and 1 MHz for continuous droplet generation. Piezoelectric dispensing technology is suitable for a wide range of material dispensing, such as liquid metals, dispersions of metal nanoparticles, electrical, and optical polymers, sealants and adhesives. The digital nature of the technology has many advantages over more conventional, subtractive and analog processes. Ink-jet-based printing requires no tooling, is noncontact and is data driven: no mask or screens are needed and the printing information is created directly from CAD or similar formats and stored digitally. It is also environmentally friendly being an additive process, generating little waste. In this chapter it will be shown how ink-jet printing could be used to create electrodes, deposit electrocatalyst, print polymer electrolyte membranes and apply sealants either as glass frits or adhesives. Some of these applications have been demonstrated for fuel cells and others have been used in related microelectronic devices.

HISTORY

Over the past almost half a century ink-jet printing technologies have been developed for low-cost, high quality office and home printers, for industrial marking applications, and for novel applications such as printing displays based on light emitting polymers, printing drug loaded polymers onto vascular stents, and printing conductive lines as current collectors on photovoltaic cells. Small office and home office (SOHO) market segment has been dominated by color ink-jet printers in last several years. Most packaged consumables such as snack foods, soft drinks, and drugs have manufacturing date, expiry date and lot codes printed on them using industrial ink-jet printers. Another industry that widely uses ink-jet printers is packaging by printing bar codes on card board boxes, enabling automated inventory management. But the fastest growing segment of the ink-jet printing industry is the large format printers that are utilized in printing low-cost, quick turnaround signs.

As early as late 1960s [1], the potential of ink-jet printing technology as a digital, high-resolution, noncontact, direct-write material deposition method was recognized. During the 1980s, electronic manufacturing, medical diagnostics, solid free-form applications were the drivers behind initial developments of ink-jet technology for novel applications. In last decade, emergence of hand held devices, human implantable devices, micro-electromechanical systems (MEMS) devices have been the driving forces in the adoption of ink-jet printing technology.

This chapter is intended to provide a broad overview of ink-jet printing technology as a low-cost, direct-write printing technology for mass production. We will start with a general description of ink-jet printing technology, the physics behind droplet generation methods, and important characteristics of these methods. This will be followed by the discussion of fluid property requirements, pattern printing (i.e., fluid/substrate interaction), and throughput. Specific examples of ink-jet deposition in electronics application will then be discussed in detail. Finally, a discussion of commercially available ink-jet systems and future trends will conclude the chapter.

BACKGROUND ON INK-JET TECHNOLOGY

Continuous Mode Ink-Jet Technology

The phenomena of uniform drop formation from a stream of liquid issuing from on orifice were noted as early as 1833 by Savart [2] and described mathematically by Lord Raleigh [3,4] and Weber [5]. In this type of system that is based on their observations, fluid under pressure issues from an orifice, typically 40–80 μm in diameter, and breaks up into uniform drops by the amplification of capillary waves induced onto the jet, usually by an electromechanical device that causes the pressure oscillations to propagate through the fluid. The drops break off from the jet in the presence of an electrostatic field called the charging filed, and thus acquire an electrostatic charge. The charged drops are directed to their desired location – either the catcher or one of several locations on the substrate – by another electrostatic field, the deflection field. This type of system is generally referred to as "continuous" because the drops are continuously generated and their trajectories are varied by the amount of charge applied. A fairly extensive theoretical and experimental analysis of continuous ink-jet devices, particularly the process of disturbance growth on the jet that leads to drop formation, has been done [6,7]. Continuous mode ink-jet printing systems produce droplets that are approximately twice the orifice diameter of the droplet generator. Droplet generation rates for commercially available continuous mode ink-jet systems are usually in the 80–100 kHz, but systems with operating frequencies up to 1 MHz are in use. Droplet size can be as small as 20 μm in a continuous system, but 150 μm is typical. Droplets as large as 1 mm (~0.5 μl) have been observed.

Figure 11-1 shows a schematic of this of ink-jet printing system and Figure 11-2 shows a photomicrograph of 50-μm diameter jet of water issuing from a droplet generator device and breaking up due to Raleigh instability (i.e., continuous mode ink-jet) into 100-μm diameter droplets at 20,000 s^{-1}.

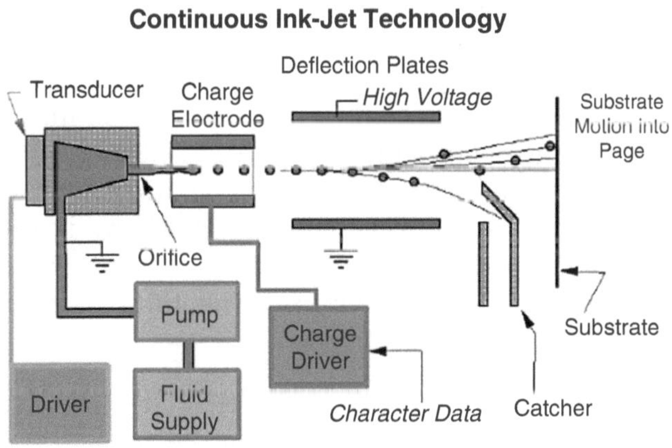

Figure 11-1. Schematic of a continuous type ink-jet printing system

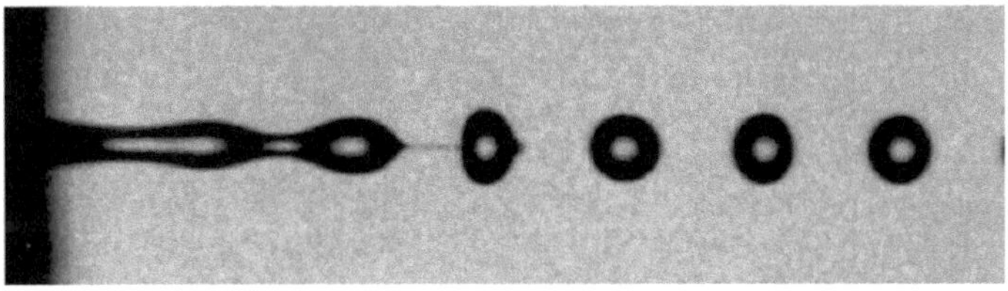

Figure 11-2. A 50-μm diameter jet of water breaking up into 100-μm droplets at 20 kHz due to Raleigh instability

Continuous mode ink-jet systems are widely used in industrial market, mainly for product labeling of food and medicines. They have high throughput capabilities, especially array continuous mode systems, and are best suited for high duty cycle applications. Few continuous ink-jet systems are multicolor (multifluid) but two-color systems are in use.

This approach is suitable for high-speed coverage of relatively large areas since drops up to 0.5 mm in diameter may be generated at rates up to 1 MHz.

Demand Mode Ink-Jet Technology

In the 1950s, Hansell observed the production of drops by electromechanically induced pressure waves [8]. In this type of system, a voltage pulse applied to a piezoelectric material that is directly or indirectly coupled to the fluid, at ambient pressure, induces a volumetric change in the fluid. This volumetric change creates pressure/velocity transients within the fluid that are directed so as to produce a drop from an orifice [9–11]. Since the voltage is applied only when the drop is needed, these types of systems are called "drop-on-demand" or "demand mode."

The "drop-on-demand" (DOD) system as shown schematically in Figure 11-3 is a more widely used and simpler approach for smaller drop (20–100 μm) and lower frequency (up to 20 kHz) printing applications. Figure 11-4 shows a MicroFab DOD ink-jet device generating 60-μm drop of butyl carbitol from a 50-μm orifice device at 2,000 s^{-1}.

The DOD device produces drops that are approximately equal to the orifice diameter of the droplet generator [12]. As shown in Figure 11-3, a DOD system is conceptually much less complex than continuous mode systems but requires the transducer to deliver three or more orders of

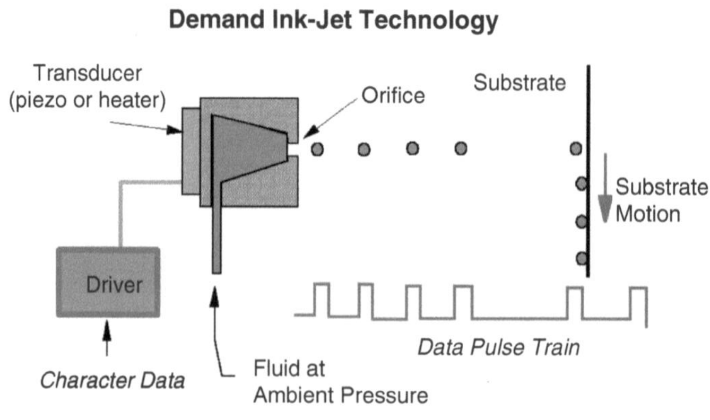

Figure 11-3. Schematic of a drop-on-demand ink-jet system

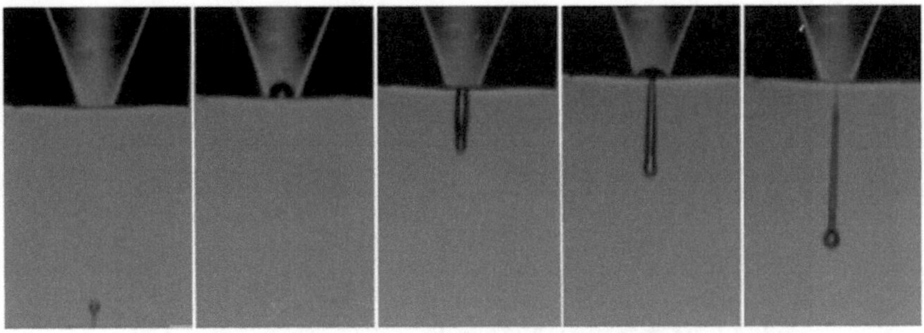

Figure 11-4. Drop-on-demand type ink-jet device generating 60-μm diameter drops at 2 kHz. Sequence from left to right show drop in-flight at different times spanning 130 μs

magnitude greater energy to produce droplets compared to continuous mode. There are many "elegant," complex array demand-mode printhead designs [13–16]. Figures 11-5 and 11-6 show 10- and 12-fluid array printheads developed by MicroFab for dispensing a range of biological fluids such as DNA, amino acids, proteins, and enzymes.

In many commercially available demand-mode ink-jet systems today, a thin-film resistor is substituted for the piezoelectric transducer. When high current is passed through this resistor, the ink in contact with it is vaporized, forming a bubble over the resistor [17]. This vapor bubble serves the same function as the piezoelectric transducer. This type of printer is referred to as a thermal ink-jet printer.

Figure 11-5. A 10-fluid array printhead

Figure 11-6. A 12-fluid array printhead

One of the characteristics of ink-jet printing technology that makes it attractive as precision fluid microdispensing technology is the repeatability of process. The images of droplets shown in Figures 11-2 and 11-4 were created by illuminating the droplets with an LED that was pulsed at the droplet generation frequency. The exposure time of the camera was ~1 s, so that images represent thousands of events superimposed on each other. The repeatability of the process results in an extremely clear image of the droplets, making it appear to be a photograph taken by a high speed camera. To further illustrate the point, Figure 11-7 shows two 60-μm diameter jets of water being caused to merge into a single droplet stream of 120-μm diameter droplets.

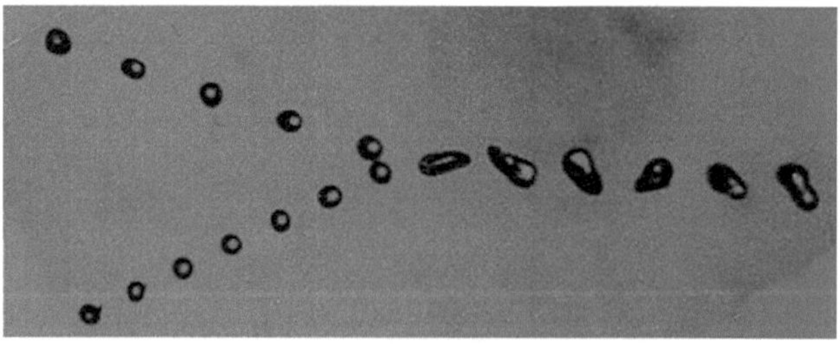

Figure 11-7. Two steams of 60-μm droplets merging into a single droplet of 120-μm at 20 kHz

FLUID REQUIREMENTS

General fluid property requirements are:

- Viscosity – Newtonian and less than 40 cP
- Surface tension – greater than 20 dynes cm^{-1}

Very low viscosities can lead to satellite formation and lack of acoustic damping but organic solvents such as methanol with viscosities less than that of water (1 cP) can be jetted. Very high surface tension presents unique challenges, but the solder discussed in "Fuel Cell Related Materials Printing" has surface tension greater than 400 dynes cm^{-1}, or roughly six times that of distilled water.

The above-mentioned properties are required *at the orifice*, if the fluid is heated or cooled. Higher viscosities are tolerated in fluid delivery system if this does not create a pressure drop that limits the desired maximum frequency. For high density fluids such as molten metals, the fluid properties should be converted to kinematic values to determine if The fluid properties are acceptable for dispensing.

The fluid requirement of Newtonian behavior is not strict but the fluid properties at the orifice flow conditions must me less than 40 cP. Thus shear thinning fluid could have shear rate viscosity much higher than 40 cP. Viscoelastic behavior causes significant performance problems by increasing the amount of deformation the fluid can withstand without breaking off from the orifice. This is illustrated by the lubricating wax shown in Figure 11-8 which forms a tail that atomizes instead of being pulled up into the main drop. Compare this with drop formation shown in Figure 11-4.

Dispersion of particles, such as metal nanoparticles, is acceptable as long as the particle/agglomerate size and density do not cause it to depart from the fluid properties range mentioned above. Particles that are greater

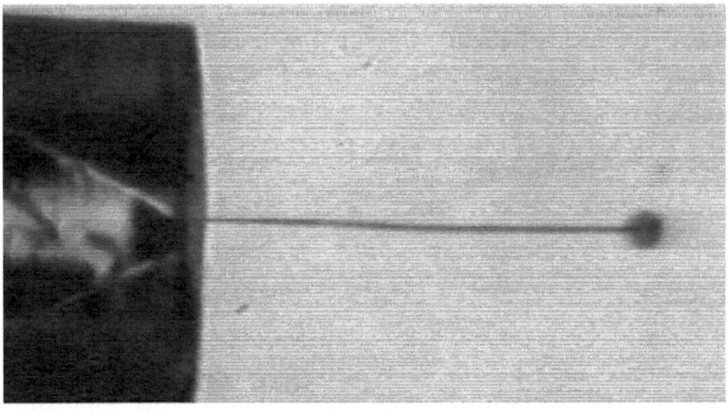

Figure 11-8. Drop formation for a non-Newtonian fluid. Compare that to Figure 11-4

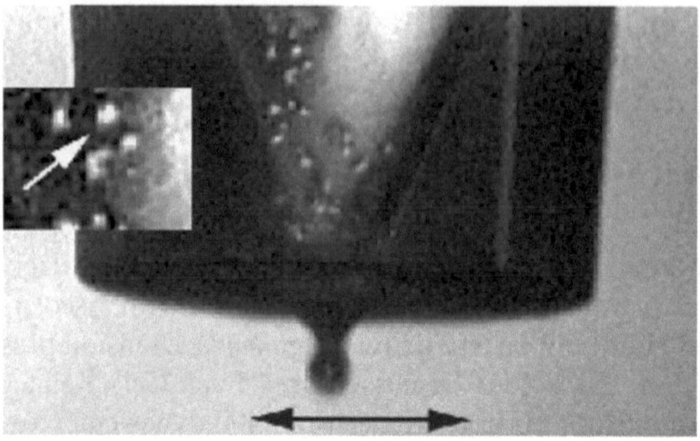

Figure 11-9. Human liver cells >10 μm being dispensed in an aqueous buffer

than 5% of the orifice diameter (e.g., human liver cell, see Figure 11-9) will cause at least some instability in drop formation behavior, but still may be acceptable in low concentration.

The "window" of fluids and suspension that can be dispensed has been stretched by heating, cooling, stirring, preoscillating, diluting, purging and wiping, and other methods. However, this operating window is unavoidably narrowed as orifice diameter decreases, frequency increases, and number of jets in an array increases.

The range of fluids that have been dispensed using ink-jet technology is quite impressive, given the fluid property restrictions just described. And that range keeps increasing as researchers come up with ways to modify materials/molecules to make them suitable for ink-jet technology. Table 11-1 lists specific materials that have been dispensed using ink-jet technology. Some of the materials on the list have wide latitude in their formulations, and hence represent a broad class of materials. Metals, oligonucleotides, and precursor formulations used for chemical synthesis (e.g., peptides, DNA) represent a class of fluids for which there is less latitude in modifying the fluid formulation to adapt to an ink-jet dispensing method.

PATTERN FORMATION: FLUID/SUBSTRATE INTERACTION

In most cases, except where it is used to meter the fluid, ink-jet deposition processes are used to produce a desired pattern of material(s) onto a substrate. The crucial part in the development of all ink-jet printing processes is the interaction between and optimization of fluid formulation, jetting parameters (waveform timings and amplitude, drop size and velocity, and

Table 11-1. Materials dispensed using ink-jet technology

Electronic/optical materials	
	Fluxes (low solids and tacky)
	Photoresists
	Dielectrics, epoxies
	Polymers (light emitting and conductive)
	Organometallics
	Cyanoacrylates
Fuel cell related materials	
	Noble metals (Pt, Pd) for electrocatalysts
	Nafion® solutions
	Metal nanoparticles (Ag, Au, Cu)
	Pb-free and Sn/Pb solders
	Sealants – glass frits or adhesives
Other particle suspensions	
	Phosphors
	Ferrites, other oxides
	Pigments
	Cells
	Teflon
	Latex spheres
Organic solvents	
	Alcohols
	Ketones
	Aliphatics
	Aromatics (xylene, toluene)
	Polar solvents (NMP, DMF, DMSO)
Biological fluids	
	Enzymes
	Biodegradable polymers (PEG, PLA)
	Lipids
	Amino acids
	Nucleic acids
	DNA
	Proteins (antibodies and antigens)
	Drug-loaded polymers
Others	
	Thermoplastics
	Thermosets
	Sol–gels
	Acrylics
	>1 M salt solutions

frequency), substrate characteristics (surface energy, roughness, cleanliness), printing grid(dpi), and printing sequence(overprinting, interleaving, printing sequence for more than one fluid) for a given application. For traditional ink-jet printing of liquid on paper, the porosity of the paper and the low viscosity of the ink represented a major challenge in the initial development of printers. Rapid spreading of the ink through the fiber can cause the dot size to become much larger(dot gain) than the drop size, causing the optical density of the spot to decrease and resulting in irregular spots that degrade the quality of lines, characters or pictures one is trying to print. The widespread acceptance of ink-jet printers was made possible by optimizing ink formulations that produced good quality prints on a wide variety of papers.

Use of ink-jet technology for printing of fuel cell components and packaging presents most of the same fluid/substrate interaction issues. The printing of Nafion®, metal catalyst (e.g., Pt), solder, electrodes, etc. requires deposition of liquid onto a nonporous substrate (similar to printing an overhead transparency). For desired resolution, it is essential to control the spreading of the material.

Phase change inks (Tektronix and Dataproducts printers) were developed for conventional ink-jet printers precisely for this reason because they solidify quickly upon impact. For fuel cell manufacturing, printing solder for interconnects is a notable example of phase change material. Figure 11-10 illustrates a solder drop that has been deposited in molten form and solidified after impact. The ridges are produced by freezing the oscillations that occur due to impact [18]. The control of the spreading due to solidification is a feature of phase change materials if one is trying to limit spreading and obtain the smallest spot for a given drop size. However, if one is trying to form a uniform layer, solidification into a bump is a problem.

At the other end of spectrum, many organic liquids, such as isopropanol, acetone, and benzene, have very low viscosity, low surface tension/low contact angle, and are volatile. These types of fluids spread quickly due to their low viscosity and ability to wet most surfaces. Depending on the application this could be either a feature or a problem just like phase change material's inability to spread. If one is trying to print a conductor using organometallic inks or create a pixel in a light emitting polymer display, spreading is definitely a problem. In many cases, surface features, such as wells commonly used in phosphor-based displays, provide a barrier to spreading, and help define the feature size.

Figure 11-10. A 60-μm solder bump deposited as a liquid onto copper using ink-jet printing technology. Left is model

In other cases, surface treatments such as application of low wetting coating or plasma cleaning are used to control the spreading by modifying the surface energy.

Volatility of solution with dissolved or suspended particles can cause operational difficulty, and drying is one of the most common failure modes with office and home ink-jet printers. In addition, for many ink-jet printing based applications volatility can cause nonuniform distribution of the solid on the substrate after drying [19], similar to the ring coffee stains form on fabric. There are many solutions to this problem: using cosolvents that are of lower volatility, covalent binding of the solid to the substrate, etc. Figure 11-11 shows a 75-μm light emitting polymer spot with nonuniform distribution of solid due to solvent volatility.

Pattern generation in its simplest form can be the selection of pixel size and spacing and then using ink-jet dispenser to fill up the desired pixel. In case of vector printing, simply selecting the spacing between drops along a line, as illustrated by the flux printing in Figure 11-12. However, there are techniques that can be used to increase the print quality: rows of spots are interleaved for high coherent errors from a single jet, multiple passes are printed over an area to increase color saturation of the printed area, and operating frequency is reduced for high-quality printing. All these techniques have applicability to printing fuel cell components and in packaging of fuel cells.

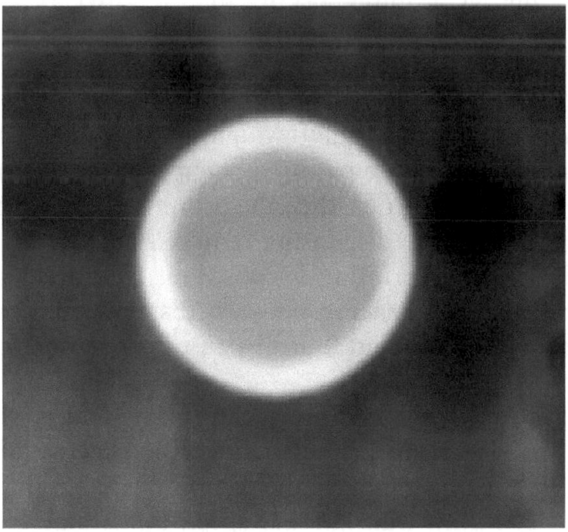

Figure 11-11. A 75-μm light emitting polymer spot with nonuniform distribution of solids due to solvent volatility

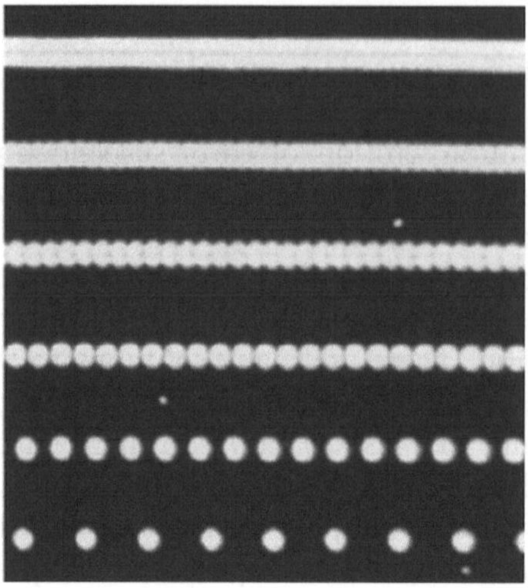

Figure 11-12. Low solids flux spots, ink-jet printed as dots and lines

THROUGHPUT CONSIDERATION

As discussed earlier, droplet generation rates for continuous and demand mode ink-jet technologies are on the order of 80 and 8 kHz, respectively. For office printers, the nominal throughput would be determined by combination of maximum droplet rate, number of orifices in the printhead, and the required printing resolution. Some of the same considerations will determine the throughput for printing fuel cell components. If multiple drops are required per spot, as is the case with some of the microoptics applications, then the printhead will normally move to a location and stop before dispensing required number of drops at that location. This is called printing-on-position. The acceleration of the moving stage(s) and the spacing between dispenses limits the effective maximum throughput to about four dispenses per second for most printing applications.

Even for print-on-fly method in which printing occurs while printhead is in motion, the way office printers works, the effective dispensing rate can be far below the maximum. If the distance between the printing sites is large (e.g., >200 μm), translational speed limits may be encountered before drop generation limits are encountered.

FUEL CELL RELATED MATERIALS PRINTING

SolderJet for Interconnect

There is a realization that the next generation packaging technology will require materials with enhanced properties and advanced manufacturing processes to address the rapidly decreasing dimensions and cost expectations [20]. Ink-jet printing is one method used to deposit materials such as solder (SnPb as well as Pb-free) for ever decreasing pitch and ball/bump size requirements, and its use has been explored by several organizations [21,22]. Continuous-mode metal jetting technology has been developed by the industry [23], academia[24,25], and national labs [26] as very high throughput method of depositing solder bumps or producing metal spheres for pick and place method (see Figure 11-13). Demand-mode solder jetting systems using both electrodynamic [27,28] and piezoelectric [29,30] transducers have been developed to overcome the drawbacks of continuous-mode systems namely long flight paths and large material usage due to waste of unused drops. The main challenge in developing demand-mode piezoelectric ink-jet devices is its ability to operate at temperatures above 200°C. Unique drive waveforms have been used in addition to selecting materials, designs, and assembly processes that are compatible with the high operating temperatures for the ink-jet devices [31].

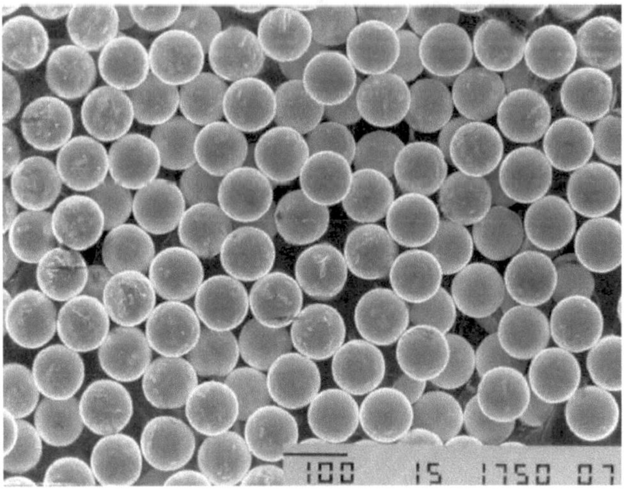

Figure 11-13. 90-μm spheres of Sn/Pb solder formed at 8 kHz and 220°C using continuous-mode ink-jet technology

Operating characteristics for solders dispensed using demand-mode ink-jet systems include solder sphere generation with diameters of 25–110 μm; drop generation rates up to 400 Hz on demand; similar rates of deposition onto pads; and operating temperatures to 320°C. The solder dispensed has been primarily eutectic 63Sn/37Pb and Pb-free 96.5Sn/3Ag/Cu, but a number of other solder alloys have been demonstrated including high lead (95Pb/5Ag), other Pb-free solders (96.5Sn/3.5Ag, indium, 52In/48Sn), and low-temperature bismuth.

Solder bumps as small as 25 μm on 35 μm have been demonstrated. Figure 11-14 also shows solder towers that can be printed by dispensing a "burst" (more than one) of drops that stack up as tower.

Figure 11-15 shows eutectic solder bumps placed onto a 18 × 18 test substrate with 100-μm diameter pads on 250 μm centers. The solder volume deposited per pad is equivalent to a drop diameter of 100 μm. The solder bumps acquire the shape shown in Figure 11-15 is a result of rapid solidification [18]. The bumps were placed at the rate of 400 s^{-1} by rastering the substrate in the horizontal direction of the rows of solder bumps.

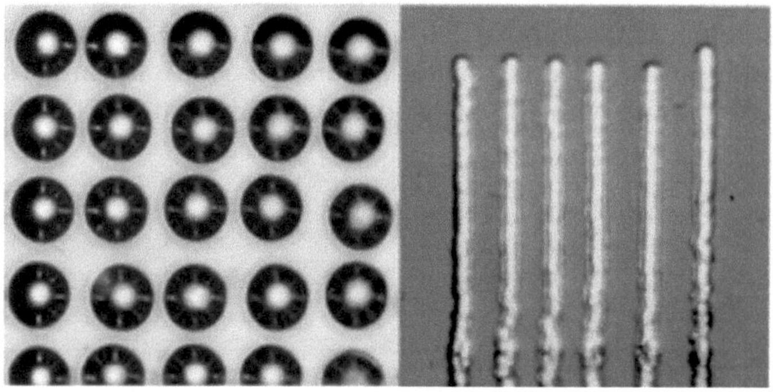

Figure 11-14. 25-μm bumps of Sn/Pb solder on 35-μm centers; *Right* is 24-μm solder towers

Figure 11-15. 100-μm solder bumps placed on 10-μm pad at 250 μm centers at 400 bumps per second

The average placement accuracy achieved in this these tests is 10 μm which is close to the accuracy limitations imposed by the motion and alignment systems of the printing platform used.

Solder jet system, being a data driven system, has the ability to control the drop volume at a given location by varying the number of drops placed at that location. In addition, unique waveforms can be employed to modulate the volume of a drop over an approximately 2:1 diameter (8:1 volume) range. Figure 11-16 shows a solder jet device generating 62-μm diameter droplets of solder at a rate of 120 s^{-1}.

A few moments later the same device is shown (see Figure 11-17) to generate much larger drops. This is achieved by applying a drive waveform that extends the drop formation process over a significantly longer time period. The diameter is increased, in this case, to 106-μm. The volume modulation using this method is continuous over the entire range of achievable volumes. This capability gives Solder Jet® technology a unique advantage to change the drop volume using software control, either for product changeover, or for application requiring variable-sized bumps onto a single substrate.

Ideally, a single drop per site is used in electronic and photonic assembly applications employing Solder Jet® technology in order to maximize throughput. However, multiple drops may be dispensed at a single site to increase the drop volume. Figure 11-18 shows an array of solder towers created by dispensing fourteen 50-μm diameter drops on top of each other

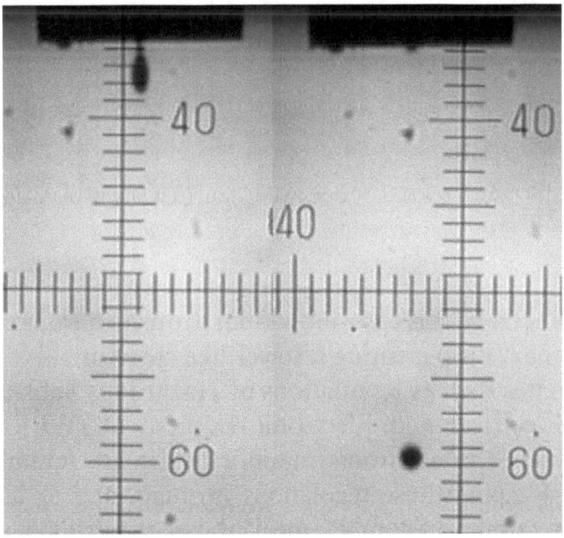

Figure 11-16. Solder droplet generation at two different times during the process. Drop size is 62-μm and drop rate is 120 Hz

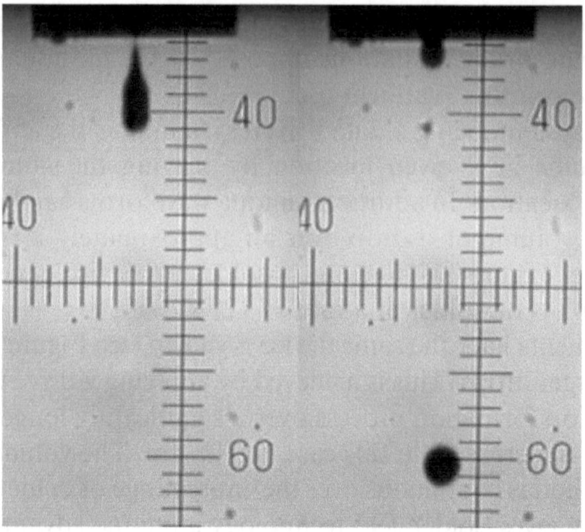

Figure 11-17. Drop formation process for same device shown in Figure 11-16, but using an extended waveform. Drop size is 10-μm and drop rate is 120 Hz

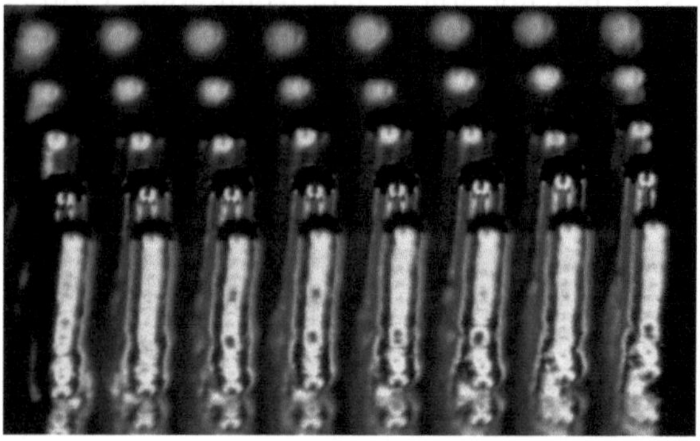

Figure 11-18. Solder towers created by dispensing fourteen 50 μm diameter droplets on top of each other at 240 Hz

at a rate of 240 Hz. In this case, individual droplets have frozen before the impact of the next drop creating a tower like structure.

With mandates such as Regulations of Hazardous Substances (RoHS), Waste from Electrical and Electronic Equipment (WEEE) companies worldwide involved in electronic manufacturing are scrambling to meet the deadlines. Keeping those regulations in mind, and to help companies involved in electronics assembly meet those regulations and deadlines MicroFab started developing Pb-free solder processes a few years ago. We are currently offering a well developed process using 96Sn/3.5Ag/Cu solder

Figure 11-19. Pb-free wafer bumping with 85-μm bumps printed on-the-fly (*top*) and in profile (*bottom*)

alloy for electrical and electronic assembly. Figure 11-19 illustrates a wafer that has been bumped using this Pb-free solder.

Embedded Passives

The need for passive components in today's electronic products has gone up dramatically. The reasons are many but principally due to increased functionality and higher performance requirements. Also, the drive to push the overall cost down, increase the reliability and miniaturization of products have motivated many OEMs to consider embedding passive components such as resistors, capacitors and inductors. A number of technologies, including ink-jet technology, are being explored to direct-write passive components. Many of the elements involved in printing passives similar to those used in printing fuel cell components.

Resistors

As a part of National Institute of Standards and Technology's Advanced Technology Project [32] resistive polymer solutions and dispersions (aqueous and organic solvent based) were dispensed to form embedded resistors on the inner layers of a multilayer printed circuit boards. The data-driven nature of ink-jet printing allows the formation of precise geometries at any location on the printed wiring board inner-layer (see Figure 11-20). Some of the resistive materials printed used carbon nanotubes as the conductive elements. The breadth of applications of this technology is only limited by the availability and environmental stability of the printing ink materials.

Capacitors and Inductors

Capacitors and inductors can be formed onto surfaces (not necessarily planar) using ink-jet printing processes by creating local 3D structures. Figures 11-21 and 11-22 schematically illustrate one method of printing capacitors and inductors. For a capacitor, bottom electrode, dielectric, and top electrode are printed successively, and this process could be repeated for a multilayer capacitors. Both the area and thickness of the dielectric could be varied to select the value of the capacitance.

Figure 11-20. Embedded polymer resistors ink-jet printed using a polymer solution. Printed resistors were <100 Ω per square. Resistors are in size ranging from 125 μm to ~1 mm

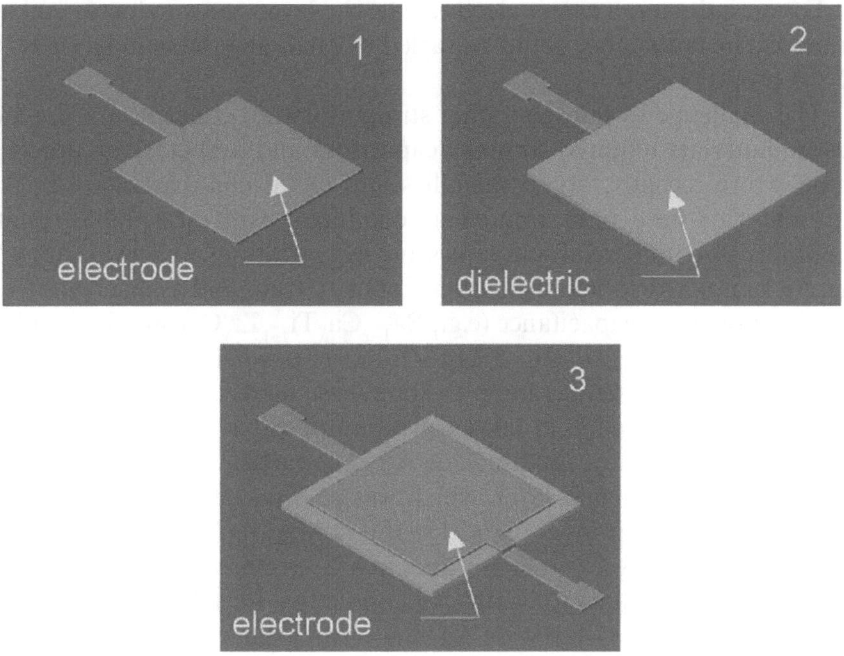

Figure 11-21. Steps for direct write inductor fabrication

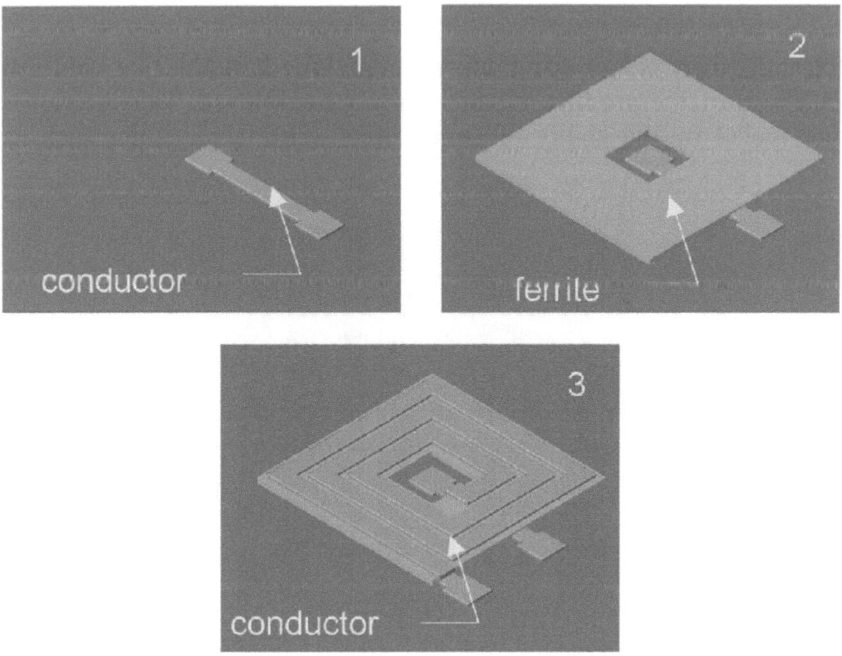

Figure 11-22. Steps for direct-write capacitor fabrication

For an inductor, a center electrode, ferrite layer and conductor coil are printed. The inductance could be varied by changing the number of turns of the printed coil.

The challenge to put this rather straightforward concept into practice lies in materials required to make capacitors and inductors of practical value. More recently, organometallics and metal nanoparticle (e.g., Ag, Au, and Cu) have been printed as conductor with acceptable results. A number of commercial companies and organizations worldwide are selling Ag nanoparticle inks including Cabot, Ulvac, and Cima Nanotech. Finally, most high-capacitance (e.g., $Ba_{1-x}Ca_xTi_{1-y}Zr_yO_3$), and high inductance materials (e.g., Ni–Zn or Mg–Zn ferrite powders) are ceramics that are sintered at high temperature. Despite these materials difficulties, some progress has been made in ink-jet printing of capacitors and inductors. Figure 11-23 illustrates 250-μm wide Ag nanoparticle lines ink-jet printed onto ferrite nanoparticle layer, which was also ink-jet printed. Since the conductor lines are printed parallel in this configuration, they would have to mate with crossover conductor lines in a layer below the ferrite in order to form a spiral conductor path. Ag conductive lines as small as 25-μm have been reported. 47-μm wide Ag nanoparticle lines printed onto a glass slide at room temperature are shown in Figure 11-24.

Batteries

Recently, there is a growing interest in making batteries that conform to the shape of the product in which they are employed. This interest is further fueled by the ability of some direct-write technologies such as

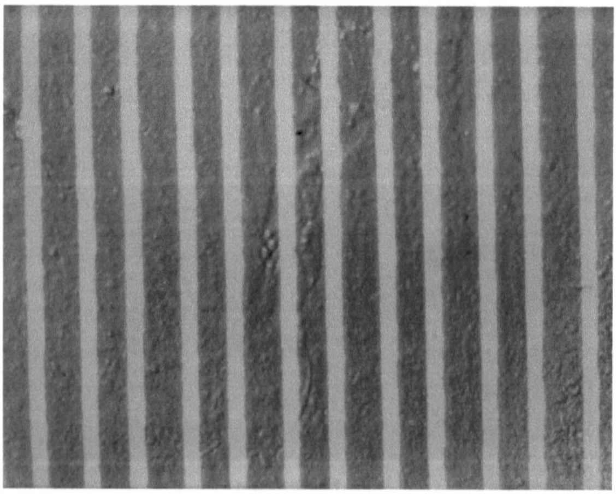

Figure 11-23. 250-μm wide Ag nanoparticle lines ink-jet printed onto ferrite nanoparticle layer (also ink-jet printed)

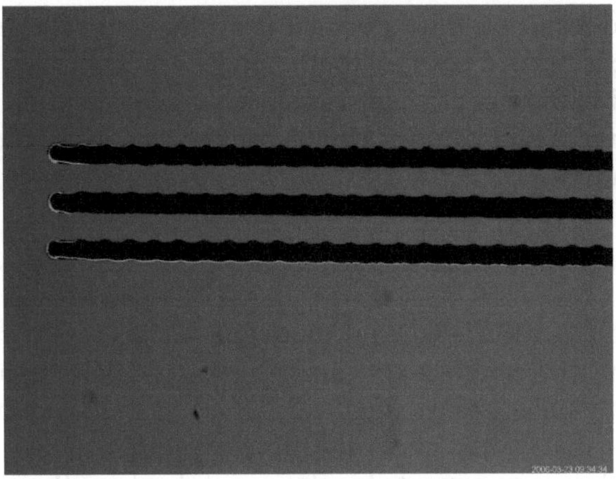

Figure 11-24. 47-μm wide silver nanoparticle lines printed on glass slide at room temperature; dot spacing 115-μm

ink-jet to print onto nonplanar surfaces, for example, helmet of a soldier, remotely piloted airplane structure, smartcards, radio frequency identification (RFID) tags. DARPA funded mesoscopic integrated conformal electronics (MICE) project was to direct write a range of electronics components including conformal batteries [33].

Batteries could be printed using ink-jet technology as illustrated in Figure 11-25. First, an organometallic or metal nanoparticle ink is printed to form metal current collectors. Next, $LiCoO_2$ is printed onto cathode

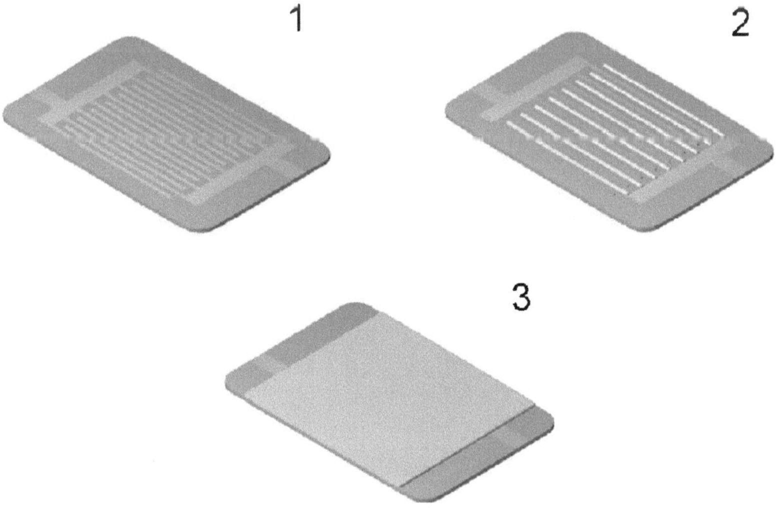

Figure 11-25. Schematics of direct-write battery fabrication

conductor and a carbon-filled polymer is printed onto the anode. Finally, the electrolyte is printed covering both anodes and cathodes. The capacity of the printed battery could be varied by changing the total area printed, and unique shapes and patterns can be printed to conform to the features of the substrate.

Spacers

The high-temperature demand mode ink-jet process used in printing UV-curing polymer microlenses can be used to create highly controlled spacers in flat panel displays. Figure 11-26 shows an example of printed spacer bumps that would meet the physical and thermal (in excess of 200°C) durability requirements for flat panel displays. Bumps as small as 25 μm diameter and 10 μm high can be created, and bumps this size or larger would span the requirements for most spacers in displays.

For spacers, the key parameter to control is the height of the deposited droplets. Height is determined by droplet volume and the degree of spreading which occurs on the substrate prior to solidification and curing. Volumetric control for demand mode ink-jet printing is usually better than 2% (range) for well behaved fluids. The spreading of deposited material may also be controlled by the use of low-wetting coatings on the target substrates to provide specific contact angles and, additionally, by varying substrate temperature to adjust fluid viscosity prior to solidification.

Figure 11-26. Array of 95-μm diameter, 34-μm high, spacer bumps in substrate plane (*top*) and in profile (*bottom*)

Adhesives

The printing of adhesives for sealing and bonding of LCD cells can be accomplished by ink-jet printing. Line patterns are printed from data files and aligned to substrate fiducials. The width and height of the cross section of each printed line is determined by the specified spacings of the deposition sites and number of droplets deposited per site, as seen in Figure 11-27.

Glass Frits

For over 40 years the semiconductor industry has used glass frits to hermetically seal ceramic cavity packages. Initially these seals were developed for high reliability MIL-STD Packaging. The sealing glasses which contained a glass frit and a binder were coated onto the lid of the package. The glass was selected to best match the thermal expansion of the package/lid combination. The binder was thermally removed before a thermo-compression seal was formed.

Currently, sealing glasses are used not only for ceramic packages but also for sealing flat panel displays, wafer level MEMS packages and sealing the chambers of solid oxide fuel cells (SOFC). Today the materials are developed to match the requirements of the specific applications. For SOFC, the seal materials contain glasses based upon the alkaline earth–zinc–silicate system, oxides to control crystallization, oxides to tailor thermal properties at high temperature and special binder systems. The glass and oxide particles in these systems can be made to less than 4 μm, with some of the oxide powders in the nanorange. At these sizes they can be put in dispersions and made into an ink-jet ink.

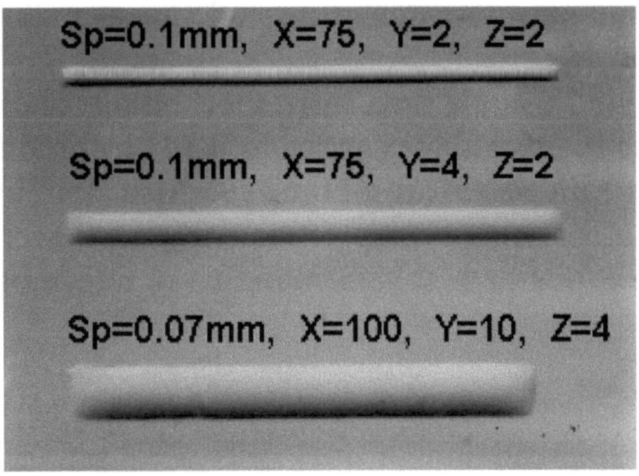

Figure 11-27. Ink-jet printed thermoset epoxy lines (top 300-μm wide), made by depositing 60-μm drops at 125°C; site spacing (Sp), number of drops in horizontal (*X*) and vertical (*Y*) directions, and number of droplets per site (*Z*) are as indicated

Today, ink-jet printing can be used to deposit a glass frit ink with significant advantages in many applications. Finer lines can be printed which reduces the real-estate used-up in the bonding process, this is especially important in wafer-level bonding. Noncontact printing enables printing on structures not possible with screen or stencil printing. In SOFC multiple single layer cells are stacked to form the final fuel cell. These ceramic layers need to be sealed to a ceramic housing to finish the assembly. Coating glass frit on the edges of each layer in the assembly can be performed with noncontact printing. The exact benefit of noncontact printing in the assembly depends upon the material systems being used and the architecture of the SOFC.

Fuel Cell Components

Fuel cells are major focus in energy and power research because they are practically emission free and quiet. In last several years, government and industry has shifted emphasis from battery-powered electric vehicle to fuel-cell powered vehicles as part of government's Clean Air Initiative.

In last few years there has been considerable amount of activity in using ink-jet technology for printing different components of a fuel cell particularly those for polymer electrolyte membrane fuel cell (PEMFC) and direct methanol fuel cell (DMFC). The various components reported to have been printed using ink-jet are as follows:

(a) Platinum electrocatalyst
(b) Fuel cell membrane using Nafion® solutions
(c) Anode and cathode

Figure 11-28 illustrates the different fuel cell components that have been reported printed by researchers in academia as well as industry using ink-jet technology.

The drive behind using a direct-write technology such as ink-jet is its potential to scale the technology up from R&D to production level once the materials and processes are optimized. Also, this technology waste little material (e.g., printing Pt catalyst) making it a low-cost alternative to conventional processes such as spin coating and screen printing.

Electro-Catalysts

A key to widespread use of fuel cells as a power source is high-performance, low-cost manufacturable electrocatalyst. Ink-jet technology has been used in library preparation for methanol fuel cell catalysts discovery at Penn State University and Illinois Institute of Technology [34].

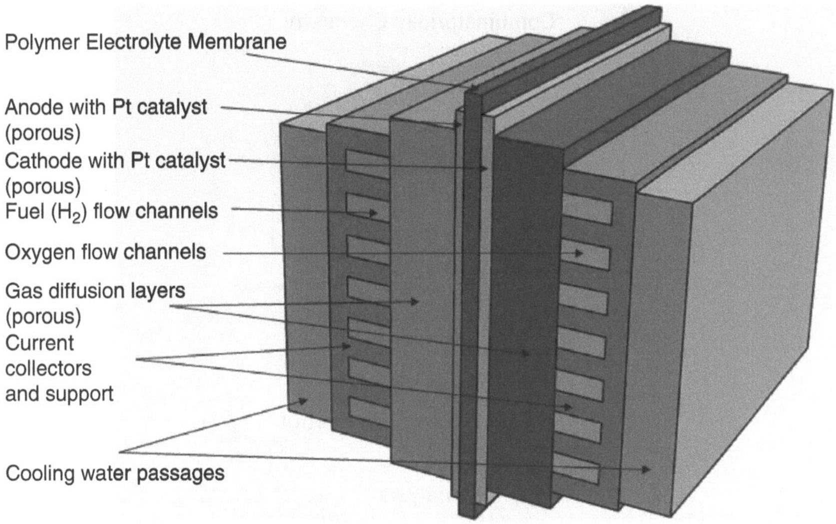

Figure 11-28. Schematics of a PEM fuel cell

Using a modified Epson ink-jet printer, they printed metal salt solutions of platinum, rhodium, osmium, ruthenium, and iridium to form 220 compositions in the phase space of ternary and quaternary alloys of these metals. They were converted to metallic form after printing the metal salt solutions and screened for performance using a pH-sensitive fluorescence indicator. This led to identification of new fuel cell catalysts $Pt_{62}Rh_{25}Os_{13}$ and $Pt_{44}Ru_{41}O_{10}Ir_5$, which provided 40% higher current densities.

Also, a group of researchers at University of Texas at Austin used ink-jet for rapid screening of non-platinum electrocatalysts such as Pd–Ti and Pd–Co–Au which show electrochemical performance similar to that found with commercial platinum catalysts [35].

Figure 11-29 shows a schematic illustrating how ink-jet printing technique can be employed to discover new and better performing electrocatalysts using combinatorial approach.

Based on certain thermodynamic guidelines, a set of materials are selected. Based on the criteria defined in "Fluid Requirements" ink-jettable inks are formulated. An ink-jet printer or jetlab-II type printing system (see Figure 11-30) is employed to dispense an array of spots onto a substrate or into wells in different combinations of metal inks. The printed samples are individually or collectively processed so that all processing variables are explored. The sample properties are measured and archived along with composition data in a database. Finally, the database is mined to find optima and make connections between the composition of the catalysts, their structure and properties.

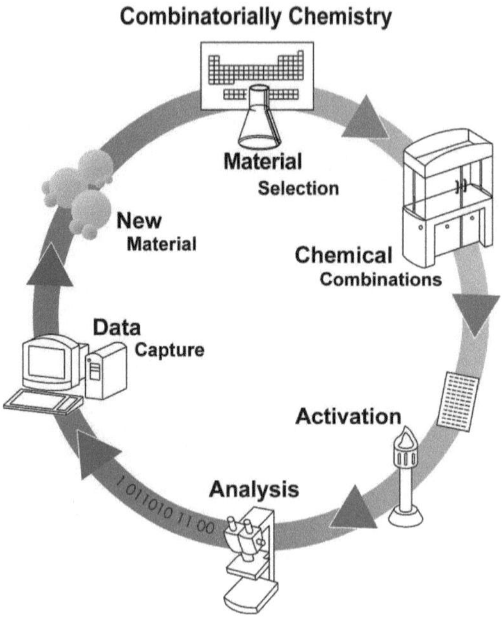

Figure 11-29. Schematics showing steps followed in combinatorial approach to finding new and improved catalysts

Figure 11-30. MicroFab's Jetlab®-II, a configurable ink-jet printing system

One Department of Energy study in 2003 estimated that platinum accounted for 10–15% of overall cost of a fuel cell production cost [36]. Any savings in printing Pt or its alloys will have direct impact on the overall production cost of fuel cells.

Polymer Electrolyte Membrane

Both aqueous and solvent based solutions of Nafion® has been printed using ink-jet methods. The solvent based solution of Nafion® has a tendency to dry around the nozzle of the jetting device. Further dilution with a solvent with higher boiling point helped to mitigate drying around the nozzle. Overall, aqueous solutions of Nafion® showed better jet stability and are more conducive to printing membranes.

Electrodes

A few researchers in industry and commercial organizations have demonstrated printing electrodes for microfuel cells with results comparable to conventional electrodes. Proprietary nature of such research efforts precludes us from sharing any more information at this time.

COMMERCIAL PRINTING SYSTEMS

In many non-ink applications of ink-jet printing technology, the printing system is viewed as a proprietary tool used to manufacture a high value added product, and part of the value usually includes the intellectual property associated with the ink-jet-based fabrication method. Thus, the availability of commercial ink-jet-based fuel cell equipment is limited compared to level of activity in applying ink-jet technology to printing fuel cells.

Commercially available ink-jet-based printing systems that are appropriate to rapid prototyping applications are given below and are divided into two categories. In section "Products", systems that are marketed to specific product areas are given. These systems will be more highly developed for their specific target applications, but will also be much less flexible or adaptable to other applications of ink-jet dispensing. Conventional ink-jet printing hardware would also fall into this category.

Under the *Research Systems* section, ink-jet-based hardware that is sold as configurable and/or appropriate to multiple applications is given. Since they are designed for flexibility and adaptability, they are, in general, less tailored to any specific application.

In addition to commercial systems, components and subsystems can be purchased from a number of companies for integration into a custom ink-jet printing platform. This includes conventional ink-jet printheads. A detailed discussion of components and subsystems is beyond the scope of this chapter.

Products

Free Form Fabrication

Solidscapes' T612 benchtop system [37], uses two demand mode piezoelectric array printheads dispensing a thermoplastic material (95–110°C melting point) to build solid models. One of the two printheads also dispenses a support material, also a thermoplastic (49–70°C), to allow it to construct overhanging structures. The minimum feature size it can achieve is 250 μm, using a 75 μm drop size. The build layer thickness (i.e., the thickness of the slice through the virtual object printed in one pass) is 13–76 μm. The build envelope or size of the largest part that can be constructed is 30 cm × 15 cm × 22 cm (*X*, *Y*, and *Z*).

The 3D Systems Thermojet Printer [38] also uses a demand mode piezoelectric array printhead dispensing a thermoplastic material to build solid models. Printing resolutions (actually addressability) of 300, 400, and 600 dot per inch (drop spacings of 85 μm, 63 μm, and 42 μm) are claimed. The maximum model size is 25 cm × 19 cm × 20 cm.

The Objet Quadra, recently developed by Objet Geometries Ltd., prints a photopolymer at 600 dots per inch in *X* and *Y*, and uses a 20-μm layer thickness. The photopolymer is cured during printing so not post-processing is required. A support material is also printed where required. The maximum model size is 27 cm × 30 cm × 20 cm.

Voxeljet Technology GmbH [39] offers a systems base on ink-jet dispensing of binder fluid onto particles, similar to the technology developed by MIT except the binder is not burned off after the part is printed. The build envelope is 85 cm × 45 cm × 50 cm. They also have a system that prints a water soluble wax that is ink-jet printed to form the boundary of a poured (in layers) wax part.

Research Printing Systems

MicroFab Technologies [40] manufactures and sells three different sizes of ink-jet-based printing systems called Jetlab®, Jetlab®-II and Jetlab®-4. Jetlab®-II and Jetlab®-4 are shown in Figures 11-30 and 11-31.

Printheads capable of printing solder, nanometal particle dispersions, polymers or bioactive materials can be incorporated in the above mentioned printing systems. Printed areas 15 × 15 cm can be addressed with accuracy of 3–10 μm and at rate up to 20 kHz onto a temperature controlled platen. The Jetlab®-II system consists of two camera vision that allows for drop-formation setup, substrate/drop alignment, and post printing inspection. It has pneumatics control for precise maintenance of meniscus at the orifice of the jetting device, drive electronics to pulse the jetting device at frequency up to 20 kHz and motion control for a 25 cm × 20 cm × 10 cm *XYZ* stage.

Microdrop GmbH [41] manufactures and sells a dispensing system that can control up to eight individual dispensers and a 20 cm × 20 cm *X–Y* stage. A vision system allows for set up of drop-formation process and observation.

GeSim GmbH [42] markets liquid-handling system for life sciences that aspirate and dispense. It can mount up to eight single-channel DOD piezoelectric dispensers, and dispense 100- and 500-pL drops (60- and 100-μm diameter drops, respectively). A vision system allows for observation and set up of the drop-formation process.

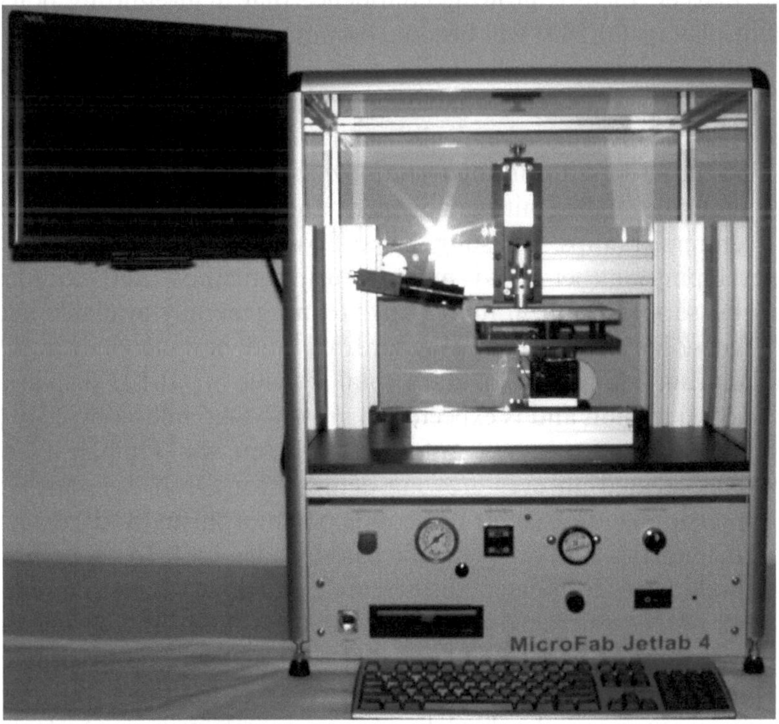

Figure 11-31. MicroFab's Jetlab®-4, a low-cost table-top ink-jet printing system

Dimatix recently introduced a table-top printer with a 16-nozzle printhead and a 20 cm × 30 cm × 5 cm *XYZ* stage. A vision system allows for observation and set up of the drop-formation process. An optional fiducial camera allows matching drop placement to previously patterned substrate [43].

FUTURE TRENDS

Given its cost/performance ratio, and more importantly its ability to print color images, ink-jet printing technology has dominated SOHO market for more than a decade now. However, current and future developments in the conventional ink-jet printing market (i.e., printing on paper) are and will be targeted to consumer market, and these developments will not affect the applicability of this technology to printing fuel cells and their components, where the "printer" is a manufacturing tool.

Recent developments in three nonconventional applications of ink-jet printing are likely to have significant impact on the overall capabilities and availability of ink-jet-based equipment for manufacturing. Clearly, in the display market, there is growing confidence that light-emitting polymer-based display (p-OLED) will become dominant technology in future displacing CRT and eventually LCD displays in many or most existing display products. It will also enable a wide range of display products that are not technically or economically viable today. There is broad acceptance that ink-jet-based printing technology is most practical for manufacturing multicolor light-emitting polymer-based displays, and many display manufacturers (Philips, Samsung, Seiko-Epson, Litrex, CDT, UDC, etc.) are developing ink-jet-based manufacturing method and equipment. Again this should result in more capable and generally available ink-jet-based technology for use in standardized and commercially available manufacturing equipment. In 2005, total revenue of OLED displays was close to $500 million and is expected to grow to $4.6 billion by 2010 [44].

The explosion in genetic information from the Human Genome Project has created an ever growing demand for DNA microarrays for gene expression studies, single nucleotide polymorphism (SNP) detection, clinical diagnostics, and other genetic based studies [45]. To date pin transfer methods have dominated the research and low volume commercial markets for DNA microarrays, and Affymetrix's DNA microarrays fabricated using light activated synthesis [46] been widely used by drug companies in their research. However, as the market for DNA microarrays becomes larger more clearly defined, companies such as Agilent and Motorola have begun to develop high volume DNA microarray production capabilities using ink-jet printing methods.

Finally, the drive to manufacture low-cost, low-capability plastic electronics has lead to the development of ink-jettable materials and processes for printing conductors, semiconductors and dielectrics. Companies like PlasticLogic have developed ink-jet-based processes and intellectual property to print display backplanes on flexible substrates for low-cost e-papers, e-books, e-maps, etc. The similar printing processes could be developed for printing RFID tags for supply chain management, bar code replacement, theft protection and contactless smart cards [47]. Over the next few years fuel cell research and manufacturing will surely benefit from ink-jet printing technology and processes developed for low-cost plastic electronics. As more tool and materials become available they will be adopted for fuel cell applications.

CONCLUSIONS

The capability of ink-jet printing technology to dispense/print a wide range of materials of interest to fuel cell manufacturing has been demonstrated. These materials include platinum electrocatalyst, Nafion® solutions as polymer electrolyte membrane, solder as interconnect, electrodes, and sealing materials. In addition to the wide range of suitable materials, the inherently data-driven nature of ink-jet printing technology makes it highly suited for fuel cell manufacturing. Using ink-jet-based methods, commercialization of products is in progress for color displays, electronics assembly, photonic elements, and DNA microarrays. In the future, both the number and type of products fabricated using ink-jet technology should increase, and the availability and capabilities of ink-jet-based prototyping and production tools should expand.

References

[1] A. Lewis, U.S. patent, 1968.

[2] F. Savart, Memoire sur la constitution des veines liquides lancees par des orifices circulaires en mince paroi. *Annales de Chimie et de Physique* 53 (1833) 337–386.

[3] J.W.S. Rayleigh, On the instability of jets. *Proceedings of the London Mathematical Society*, 10(4) (1879) 4–13.

[4] J.W.S. Rayleigh, On the capillary phenomena of jets. *Proceedings of the Royal Society of London*, 29 (1879) 71.

[5] J.W.S. Rayleigh, On the instability of a cylinder of viscous liquid under capillary force. *Philosphical Magazine*, 34 (1892) 145–154.

[6] K.C. Chaudhary, I.G. Redekopp, and T. Maxworthy, The non-linear capillary instability of a liquid jet. *Journal of Fluid Mechanics*, 96(Part II) (1979) 257–312.

[7] W.T. Pimbley, Drop formation from a liquid jet: a linear one-dimensional analysis considered as a boundary value problem. *IBM Journal of Research Development*, 29 (1984) 148–156.

[8] C.W. Hansell, Jet sprayer actuated by supersonic waves. U.S. patent 2,512,743, 1950.
[9] D.B. Bogy and F.E. Talke, Experimental and theoretical study of wave propagation phenomena in drop-on-demand ink jet devices. *IBM Journal of Research Development* 29 (1984) 314–321.
[10] J.F. Dijksman, Hydrodynamics of small tubular pumps. *Journal of Fluid Mechanics*, 139 (1984) 173–191.
[11] R.L. Adams and J. Roy, A one dimensional numerical model of a drop-on-demand ink jet. *Journal of Applied Mechanics,* 53 (1986) 193–197.
[12] D.B. Wallace, A method of characteristics model of a drop-on-demand ink-jet device using an integral method drop formation model. ASME publication 89-WA/FE-4, 1989.
[13] J.R. Pies, D.B. Wallace, and D.J. Hayes, High density ink jet printhead. U.S. patent 5,235,352, 1993.
[14] J. Roy and J.S. Moore, Drop-on-demand ink jet print head. U.S. patent 5,087,930, 1992.
[15] P.A. Hoisington, R.R. Schaffer, and K.H. Fischbeck, Ink jet array. U.S. patent 4,835,554, 1989.
[16] M. Usui and T. Katakura, Actuator for an ink jet print head of the layered type with offset linear arrays of pressure generating chamber. U.S. patent 6,033,058, 1996.
[17] J.S. Aden, J.H. Bohorquez, D.M. Collins, M.D. Crook, A. Garcia, and U.E. Hess, The third generation hp thermal inkjet printhead. *Hewlett-Packard Journal* 45(1) (1994) 41–45.
[18] J.M. Waldvogel, D. Poulikakos, D.B. Wallace, and R.M. Marusak, Transport phenonmena in picoliter size solder droplet dispensing on a composite substrate. *Journal of Heat Transfer* 118 (1996) 148–156.
[19] R.D. Deegan, O. Bakajin, T.F. DuPont, G. Huber, S.R. Nagel, and T.A. Witten, Capillary flow as the cause of ring stains from dried liquid drops. *Nature*, 389 (1997) 827–829.
[20] 1998 National Electronics Manufacturers Initiative Roadmap, 1998. http://www.inemi.org
[21] N. Van Veen and D. Schwarzbach, Solderjetting, a software driven technology for maskless waferbumping. *Proceedings, IMAPS International Symposium on Microelectronics*, Chicago, 1999, pp. 154–159.
[22] C.W. Argento, T. Flynn, and C. Demers, Next generation solder jetted wafer bumping for very fine pitch flip chip technology applications and beyond. *Proceedings, IMAPS International Symposium on Microelectronics*, Chicago, 1999, pp. 160–165.
[23] R. Godin, S. Pearson, and R. Lasky, A novel process for solder deposition. *SMT Magazine*, 1 (1997) 66–68.
[24] E.P. Muntz, M. Orme, G. Pham-Van-Diep, and R. Godin, An analysis of precision, fly-through solder jet printing for DCA components. *Proceedings, ISHM '97*, Philadelphia, 1997, pp. 671–680.
[25] J. Priest, E. Jacobs, C. Smith, P. DuBois, B. Holt, and B. Hammerschlag, Liquid metal-jetting technology: application issues for hybrid technology, *International Journal of Microcircuits & Electronic Packaging*, 15(3) (1994) 219–227.
[26] C. Hendricks, Inertial confinement fusion targets. *Proceedings of the 2nd International Colloquium on Drops and Bubbles*. NASA-CR-168848, JPL 82-7. 88-93, 1982.
[27] T.M. Smith and R.E. Winstead, Electrodynamic pump for dispensing molten solder. U.S. patent 5,377,961, 1995.
[28] T. Schiesser, E. Menard, T. Smith, and J. Akin, Microdynamic solder pump: drop on demand eutectic SnPb solder dispensing device. *Proceedings, Surface Mount International*, San Jose, CA, 1994, pp. 501–509.
[29] D.J. Hayes, D.B. Wallace, M.T. Boldman, and R.M. Marusak, Picoliter solder droplet dispensing. *Microcircuits and Electronic Packaging*, 16(3) (1993) 173–180.

[30] D.J. Hayes, W.R. Cox, and M.E. Grove, Microjet printing of polymers and solder for electronics manufacturing. *Journal of Electronics Manufacturing*, 8(3 and 4) (1998) 209–216.
[31] D.B. Wallace, Method and apparatus for forming microdroplets of liquids at elevated temperatures. U.S. patent 5,415,679, 1995.
[32] Advanced Embedded Passives Technology final report. http://aept.ncms.org/about.htm
[33] http://www.darpa.mil/dso/trans/mice.htm
[34] E. Reddington, A. Sapienza, B. Gurau, R. Viswanathan, S. Sarangapani, E.S. Smotkin, T.E. Mallouk, Combinatorial electrochemistry: a highly parallel, optical screening method for discovery of better electrocatalysts. *Science* 280(5370) (1998) 1735–1737.
[35] J.L. Fernández, V. Raghuveer, A. Manthiram, and A.J. Bard, Pd–Ti and Pd–Co–Au Electrocatalysts as a replacement for platinum for oxygen reduction in proton exchange membrane fuel cells. *Journal of American Chemical Society*, 127(38) (2005) 13100–13101.
[36] http://www1.eere.energy.gov/hydrogenandfuelcells/pdfs/tiax_platinum.pdf
[37] http://www.solid-scape.com/t612.html
[38] 3D Systems, http://www.3dsystems.com
[39] Voxeljet Technology GmbH http://generis.de/voxeljet_en/
[40] MicroFab Technologies, Inc. http://www.microfab.com
[41] http://www.microdrop.com
[42] http://gesim.de/home.htm
[43] http://www.dimatix.com
[44] http://displaysearch.com
[45] K. Aschheim, Gene detection by array. *Nature Biotechnology* 18 (2000) 1129.
[46] S.P.A. Fodor, J.L. Read, M.C. Pirrung, L. Stryer, A. Tsai Lu, and D. Solas, Light-directed, spatially addressable parallel chemical synthesis. *Science* 251 (1991) 767–773.
[47] http://www.plasticlogic.com

INDEX

MIX
Papier aus verantwortungsvollen Quellen
Paper from responsible sources
FSC® C105338
www.fsc.org

If you have any concerns about our products,
you can contact us on
ProductSafety@springernature.com

In case Publisher is established outside the EU,
the EU authorized representative is:
Springer Nature Customer Service Center GmbH
Europaplatz 3, 69115 Heidelberg, Germany

Printed by Libri Plureos GmbH
in Hamburg, Germany